FROM DUALISM TO UNITY IN QUANTUM PHYSICS

FROM DUALISM TO UNITY IN QUANTUM PHYSICS

BY

ALFRED LANDÉ

Emeritus Professor of Physics,
Ohio State University

CAMBRIDGE
AT THE UNIVERSITY PRESS
1960

The text of this monograph rests on articles published in *The Physical Review* (U.S.A.); *American Journal of Physics* (U.S.A.); *Philosophy of Science Journal* (U.S.A.); *Philosophy of Science Journal* (British); *Endeavour* (British); *Mind* (British); *Journal de Physique* (France); *Physikal. Zeitschrift* (Germany); *Naturwissenschaften* (Germany); *Philosophia Naturalis* (Germany); *Current Science* (India); *Dialectica* (Switzerland). It is a further development of the author's *Foundations of Quantum Theory, a Study in Continuity and Symmetry*, Yale and Oxford University Presses, 1955.

CAMBRIDGE
UNIVERSITY PRESS

University Printing House, Cambridge CB2 8BS, United Kingdom

Cambridge University Press is part of the University of Cambridge.

It furthers the University's mission by disseminating knowledge in the pursuit of education, learning and research at the highest international levels of excellence.

www.cambridge.org
Information on this title: www.cambridge.org/9781316509760

First published 1960
First paperback edition 2015

A catalogue record for this publication is available from the British Library

ISBN 978-1-316-50976-0 Paperback

CONTENTS

PREFACE

Albert Einstein once remarked: 'If you want to find out about the methods of the theoretical physicists . . . do not listen to their words; fix your attention on their deeds.' With all reverence to the magnificent deeds of the theorists, they have turned to words at a far too early stage of development when deeds were still in order. This holds in particular for the doctrine that a fundamental wave-particle duality is an immanent feature of the microcosm which must be accepted at face value without permitting any further explanation. To my mind, this doctrine relies too much on the policy: if you cannot explain it, call it a principle; then defend it as fundamental and absolutely irreducible, so that speaking of the unsolved 'riddle of duality' from here on becomes the mark of naïveté if not of heresy. I am the last to deprecate the interplay of corpuscular and undulatory phenomena bridged by complementarity as a valuable heuristic viewpoint. But I submit that duality as a fundamental principle of two 'pictures' of equal rank suffered a fatal blow already thirty years ago when Max Born, in his statistical interpretation, told us in simple and unambiguous terms that matter still consists of discrete particles, all wave-like appearances notwithstanding. This first step towards a unitary quantum mechanics of particles remained a torso, however, since no reasons were given as to *why* matter particles (and field oscillators, etc., as 'particles of radiation') should obey the quantum rules of probability interference and h-dominated periodicity, so that we might gain a better understanding of the rather baffling quantum phenomena. In order to fill this gap, the present study reopens the question (which had caused so many headaches to the pioneers, but has recently almost been forgotten): why do particles, in their statistical behaviour, obey wave-like laws? An answer is obtained on physical grounds alone; philosophical arguments as to why we should be content with *not* answering the question can thus be spared.

Chapter I deals with the age-old controversy between determinism and chance. Instead of accepting chance law as an irrefutable fact as such, statistical distribution is linked to the postulate of cause-effect continuity in the macroscopic domain, for example in thermodynamics; it is an attempt to give sufficient reasons for the lack of sufficient causes. Chapter II establishes an operational definition of the various 'states' of an object in general. It turns out that the concept of a 'fractional degree of equality' between two states is quite fundamental for obtaining a well-ordered schema of all 'states' of an object in the sense of the quantum theory. Chapter III demonstrates that the wave-like law of probability interference is not an oddity of nature dictated by a separate 'principle' of wave-particle duality; it is a mathematical matter of course under the simple postulate that the probability tables which connect sets of states are, in their turn, connected by a general self-consistent metric law, rather than presenting a chaos. The recognition that ψ-laws are but metric relations between statistical tables ought to dispel any idea that Schrödinger's ψ-waves have physical existence of their own, that they may expand, shrink, or 'pilot' observable events. Chapter IV gives a deduction, based on familiar theorems of classical statistical mechanics, of the most significant feature of quantum mechanics which is, the wave-like connection between co-ordinates and momenta. Chapter V criticizes the predominant quantum philosophy and language and pleads for a return to common sense from quite unnecessary profundities. Altogether the book seeks to establish a kind of *quantum genetics* in place of the faithful acceptance of the quantum rules merely as instruments of description and prediction.

Spring, 1960 A.L.

INTRODUCTION

When dealing with the foundations of quantum mechanics one must clearly distinguish between the general and the special theory. General quantum theory is concerned with the formal rules and calculation methods, with $E = h\nu$, $p = h/\lambda$, and their modern counterparts, namely Born's qp-commutation rule and Schrödinger's p-operator rule—irrespective of whether these rules are applied to electrons, neutrons, or mesons, to oscillating matter particles or to field vibrations. Likewise, the much discussed wave-particle dualism refers to electrons versus electronic Schrödinger waves, to mesons versus Yukawa waves, and so forth. Most physicists take the general quantum rules for granted and regard them as mathematical formulations of the principles of duality, uncertainty, and complementarity, taken as fundamental and irreducible to anything more elementary. The big problem today is the special quantum theory, not yet existing, whose aim is the explanation of the variety of 'fundamental' particles and their fields. Of this special problem I shall have nothing to say. My topic is rather antiquated; it goes back more than half a century to the time when the opposite situation prevailed, that is when electrons and protons were taken for granted as the ultimate building stones of matter, whereas the formal quantization rules were felt as presenting a great enigma. Physicists wanted to *understand* the quantum rules in terms of familiar features of mechanics. In particular during the first decade after Planck's discovery, some of the keenest minds tried very hard to detect some flaw in the proof of the classical energy equipartition theorem which might leave room for discrete energy levels. It was a futile approach; if one wishes to *explain* the quantum rules he must start from a statistical basis right away, forget about classical mechanics, and be glad if the latter comes out at least as an approximation.

But then, what does it mean 'explaining' the quantum rules? For illustration look at relativity theory and imagine for a moment that the mass of a body had been found experimentally to depend on the velocity, furthermore that there is a contraction of travelling rods and a time dilatation of travelling clocks, and finally that there is a mass-energy relation, $E = mc^2$. It certainly would be pleasant to find that all these perplexing results are consequences of *general* principles valid for both mechanics and optics. One may of course try to interpret the same experimental results in terms of an invisible *ad hoc* invented ether. But only Einstein's explanation, that is his derivation of the various relativistic effects from fundamental postulates of symmetry and invariance makes us *understand* them as necessities rather than as oddities. In the quantum theory, however, we are still in the stage of accepting $E = h\nu$ and the other quantum rules at face value without further explanation. Self-invented quantum potentials and the like do not help. Nor does the word 'duality' explain anything; it is a purely descriptive term. Even worse, the road to a better understanding has been blocked for thirty years by the creed that duality itself represents the bottom of theoretical analysis, and that 'we must understand that there is nothing further to be understood' in quantum theory. I beg to challenge this view, and I shall try to show that the amazing quantum formalism, the Born commutation rule, the wave-like qp-periodicity, the interference law for an amplitude function ψ, and all the rest can be deduced from, or explained on the grounds of, simple and general postulates of invariance, symmetry, etc.—just as $E = mc^2$ can be derived from Einstein's postulates.

However, before presenting the desired derivation of the quantum rules, I may be permitted shortly to criticize the customary approach together with the much hailed quantum philosophy of the Schools. Here we are, faced with the interesting problem as to *why*, on which general grounds, particles (and fields) behave in a wave-like statistical manner controlled by complex-imaginary ψ-functions, non-commutative multiplication, etc. But instead of helping us to solve these 'quantum riddles' by the means and methods of theoretical physics (as Einstein did for the Michelson-Morley riddle), the dominant School gives us stones of philosophical argument rather than the bread of physical explanation. First, our curiosity is blunted by the declaration that there can be no explanation anyway, that economical description is all we may strive for. Next, if there are two conflicting descriptions, one of waves and one of particles, this is supposed to bring out the philosophical lesson that we should not bother about an alleged 'real' world, but be content with pictures and other constructs of our mind. No wonder if Sir James Jeans([1]), after having studied Bohr and Heisenberg, arrived at the triumphant conclusion that nature consists of waves of knowledge, or of absence of knowledge, in our own minds.

All this wisdom has taken its starting-point from the empirical fact that position and momentum, q and p of a particle cannot be measured with such accuracy as to permit accurate *predictions* of the future q's and p's, according to the Heisenberg uncertainty rule. Next, this qp-uncertainty is supposed to mean that an electron *is* neither a particle nor *is* it a wave—a rather odd statement in view of the fact that one still can *count* electrons by their discrete charges, rest masses, and spins, even though electronic particles have some unexpected new qualities such as qp-incompatibility. Yet no physicist in his private thought regards an electron as anything but a discrete particle. The motive for publicly denying that an electron is a particle may be seen in the irrational desire to picture it *also* as a wave and to support the very dubious statement that 'an electron sometimes behaves like a concentrated particle, sometimes like a wave filling the whole space'. Actually, an electron always behaves exactly as a particle ought to behave. It is a myth that it sometimes misbehaves.

The latter assertion had its origin in the diffraction phenomena of electronic rays through two slits, where one may ask: 'how can an electron passing through one slit "know" of the presence of the other slit so that it will contribute to a diffraction pattern clearly depending on both slits?' This apparent *telepathy* has bothered Einstein to such an extent that he regarded quantum mechanics as an incomplete theory. And for Bohr and Heisenberg the same telepathy has been the starting-point of their *dualism* according to which 'I must not say "an electron is a particle" or "it is a wave", but "I decide by the disposition of my experiments in which of the two ways it manifests itself".' (von Weizsäcker)([2])

The situation is sometimes characterized by the word 'wavicle', to which there are serious objections, however. First, a wave has intensity and phase; but where has a single electron ever displayed an intensity and phase? Both occur only in the statistical display of many electrons. Second, a wave is spread out in space; but in diffraction it is the crystal or screen, and not the diffracted electron, which fills a large portion of space. It is not even *as though* an electron were spread out like a wave, since a large body naturally reacts as a whole to incident small particles. And the mechanical reasons for periodic diffraction patterns have been known since 1923 when W. Duane developed the first corpuscular interpretation of diffraction through a crystal, followed by Ehrenfest-Epstein's theory for screens with slits. However, neither Duane nor his successors provided us with an *explanation* of the quantum rule, $p = h/\lambda$, which they took for granted. Since an explanation, that is a reduction to more elementary postulates, is now at hand (Chapter IV), the *wavicle dualism* has become an anachronism. Shifting the ground to the *neo-dualism* of Klein, Jordan, and Wigner (section 29) or suggesting a face-saving *neo-neo-dualism* (Retrospect) will hardly discredit the notion that particles are concrete *real things*, in contrast to ψ-waves as abstractions, as mathematical descriptions of statistical qualities.

Theorists, however, take pride in putting the word 'real' always in quotation marks, in order to indicate that all things, from trees to electrons and electric fields, are constructs of our minds, are pictures. Very well! But then they ought to put ψ-waves in double quotation marks as pictures within the particle picture. My suggestion is to indulge in such luxury only in philosophical discourse, and to speak in the laboratory and in popular addresses without qualms of trees, electrons, and electric fields as *real* (kickable: Dr Johnson) things, and to put only the (non-kickable) ψ-waves as mental abstractions in quotation marks. Then, however, it is a violation of elementary logic to proclaim a *duality* between a real thing and an abstraction, as though they were on a level of parity.

But once the dualistic doctrine had spread that individual electrons sometimes misbehave as though filling the whole space like a wave (instead of correctly saying that the dispersion of *many* electrons obeys a wave-like statistical law), there was no end to further misunderstandings. If a particle is not always a particle, it means that 'we must not think in objects any more', that 'there is a collapse of the category of substance', and 'we must adapt our logic to the new situation' (von Weizsäcker's *World View of Physics*). I for my part refuse to adapt my logic, and stop thinking in objects every time I see electrons forming a periodic statistical arrangement, after having passed through a periodic crystal.

It is true that dualism, with neutrality towards two conflicting views of the real constitution of matter, had a certain shock-absorbing merit thirty years ago in reaction to the discovery of matter ray diffraction and its successful description by the wave equation. However, the question of which is *real* and which is *apparent* came up immediately. First Schrödinger proposed that particles are illusions, in reality produced by high crests of waves in a continuous substratum. This *unitary wave interpretation* of particle appearances proved untenable, however, at least in its primitive form. Then Born introduced the *unitary particle interpretation*: wave-like phenomena are actually produced by the statistical co-operation of particles. But if Born is right—and hardly any physicist doubts it because of the observed statistical build-up of interference patterns—then it is a violation of the simplest rule of orderly thinking to insist on a dualistic opposition of a *thing* (for example an electron) versus one of its many qualities (a wave-like statistical distribution of many electrons) under special circumstances. Yet, the peaceful co-existence of two absolutely incompatible ideologies, namely Born's unitary statistical particle interpretation and Bohr-Heisenberg's dualism with particles and waves on a par, has enveloped the quantum theory in self-contradictions and pseudo-problems which have led to endless discussion. Typical in this respect is *The Strife about Complementarity* in which one of Bohr's disciples(4) not too successfully(5) tries to expound his Master's Voice. Typical also are the seven different interpretations of the most common symbol of quantum mechanics, namely the letter ψ, listed in section 33.

But how is it possible that there is so much disagreement that 'The Interpretation of Quantum Mechanics' is one of the favourite topics of discussion today? The reason, as I see it, is first, the idea that quantum theory deals with expectations of observers, leading to the notion that a 'state function' contracts when an observer acquires knowledge (section 33). This subjective trend began already in classical statistical thermodynamics when the macroscopic entropy was said to reflect the degree of microscopic *information* possessed by an observer, rather than the degree of molecular disorder irrespective of what an observer knows or thinks he knows.

The second ground for dispute on the interpretation of the quantum theory is the present *scholastic approach* to the theory, resting on the maxim: 'In the Beginning was the Word, or the Cabbalistic Symbol', and let the wise men afterwards illuminate its true meaning. Indeed, as far as a systematic approach is taken at all, for example von Neumann's, it starts out from the mathematical symbolism as the primary material, and only later inquires: what can be the physical meaning of this perplexing formalism? Does it describe processes within a continuous medium, or a statistical distribution of sudden events? Does it allow a deterministic interpretation, perhaps with the admission of hidden parameters? Does it concern physical states of objects, or does it describe states of mind in an observer's brain? Aldous Huxley somewhere wrote about medieval thought: 'words did not stand for things; things stood for pre-existing words. This is a pitfall which in the natural sciences we have learned to avoid.' Unfortunately, the dispute about the meaning of ψ does not confirm Huxley's optimistic view. For this reason I think that quantum mechanics ought to take a new start, not from various interpretations of *mathematical* formulæ involving enigmatic symbols and resulting in various pictures and mental constructs, but from simple *physical* axioms or postulates of a general morphological character, similar to those of relativity theory. In other words, we ought to try some measure of 'world building' (Eddington). But instead of beginning at the top with cosmological considerations, fixing constants, I shall start from the bottom with simple and general aspects about the formal connection between statistical data under simple postulates of symmetry, invariance, and the like, and then show that their combination leads, by pure reasoning, to further physical consequences which agree with the observations of quantum physics and are described by the well-known formalism of the quantum theory. Born's unitary statistical particle interpretation can thus be supplemented by an *explanation* of the wave-like statistical behaviour of particles on the grounds of 'principles nearly all so evident that one only needs to understand them in order to assent to them' (Descartes).

NOTE

Superior figures in the text thus (1) refer the reader to the References on page 111.

CHAPTER I

CAUSALITY, CHANCE, CONTINUITY

'There is no cause or effect in nature; nature has but an individual existence, nature simply *is*. Recurrences of like cases exist but in the abstraction which we perform for the purpose of mentally reproducing the facts.'

ERNST MACH

1. *Hard and soft determinism*

Before passing judgment upon determinism and chance in the age of modern science, it is necessary to clarify the terms. Universal determinism may be defined as the Laplacian doctrine that the world as a whole runs off like a tight-fitting clockwork, with its present state being connected by an unambiguous relation with every past and future state. A superior spirit viewing the whole show from outside would be able, after a short glance now, to predict and postdict the history of the world from the infinite past to the infinite future. In the space-time picture of relativity theory nothing ever happens, the world lines of all mass points being fixed relative to one another once and for all. Laplace's doctrine is often equated to the empirical law of causality; in fact, it disposes of causes and effects, unless 'cause' is taken as synonymous with 'before' and 'effect' with 'after'. The world as a whole simply *is*, and the empirical content of the *causal law* of 'equal causes, equal effects' applied to Laplace's world is *nil* since no allowance is made for finding out whether variations of cause produce variations of effect which, by the way, ought to change not only the future but the past as well. These and other fallacies, pointed out in particular by K. R. Popper(1), reveal the deterministic doctrine as being far removed from actual or possible experience; it is an academic dream.

Determinism, or rather causality, acquires a testable meaning, as a law of nature which may be right or wrong, only when the world as a whole is replaced by a section of the world as the *object*, and an outside *observer* who can modify the external conditions for the object at his will, study the ensuing changes in the state of the object, and regard them as effects of his own doing. In order to eliminate the subjective element, the observer may insert, as an intermediary between himself and the object *a*, a second body *b* as an *instrument* which interacts with the object *a*. However, a given complex, $a+b$, simply *is* without distinction of causes and effects. The fire *b* becomes the cause of the pot *a* boiling only as the extension of someone's hand deliberately starting *b* and producing an effect *a* on the pot from outside. Ernest Nagel([2]) remarks: 'the question of determinism is best construed as dealing with a procedure for the conduct of a cognitive inquiry, rather than with a thesis concerning the constitution of the world'. Cognitive inquiry, however, refers to the manual and mental activity of a *subject* who plans and performs experiments at his will. In a similar vein Max Black([3]) points out that 'cause' acquires a meaning only as a conscious and voluntary act of a *subject* 'making something happen' to an object. And 'anything having the tendency to show that the agent was not acting freely but responding to constraint . . . would immediately show . . . that *he* was not the cause but merely an instrument or an intermediary between the true cause and its effects'. As strange as it may sound to the physicist (whose thinking and language is tainted with metaphysical notions): a definition of 'cause' other than in terms of a freely chosen act applied from outside to an object, would deflate causality to a mere temporal relation. Our almost instinctive trust in the law of *sufficient causation* then seems to have its root in our feeling of *free will*, of being free agents acting on the unfree rest of the world – a somewhat paradoxical yet almost inevitable psycho-physical theory. The same idea is expressed by von Weizsäcker([4]): 'The human being can by arbitrary acts influence the future, but not the past. Since we possess no causal description of human acts of will, a one-sided causal chain running into the future begins for us with every such act'. The physicist von Weizsäcker and the philosopher Black independently arrive at the same conclusion that the only 'causes' are personal acts of will.

The aim of the physical scientist has always been the elimination of the personal factor. He replaces his eyes and ears by a measuring instrument brought in contact with the object. Then he interprets the reaction of the instrument as indicating a certain *state* of the object according to his and his colleague's theory. He finds that the recorded state of the object depends on its preparation through previous contact with another object, which in turn may have been prepared by contact with a third object, and so forth. All this involves voluntary acts of composition and separation; hence 'causes' are reduced to deliberate acts of *preparation of a state* (Margenau)([5]). For example, the preparation may be the lifting of a stone to a certain height and releasing it there; the recording instrument may be the ground hit at this or that 'measured' place.

To the heirs of Bacon and Descartes it may seem self-evident that one and the same 'cause' (= preparation of a state) will always have the same 'effect' (= record on a certain instrument), and that apparent deviations from this rule ought to be traceable to concealed deviations of cause. The 'ought' is suggested by a preconceived deterministic ideology, however. The greatest challenge to the doctrine of 'equal causes (= preparations), equal effects' is the statistical dispersion of effects in *games of chance*, be they dice games, series of atomic tests, or the games which insurance companies play with their clients. Suppose that an experimentally prepared state A, duplicated as closely again and again, is followed by different records, sometimes a, sometimes a', sometimes a'', etc., in a certain statistical frequency ratio (1 a).

$$A \begin{array}{l} \nearrow a \\ \rightarrow a' \\ \searrow a'' \end{array} \qquad (1a)$$

We now ask:

(1) Is there reason to hope that one may explain, that is reduce the statistical frequency ratio of $a{:}a'{:}a''{:}\,..$ to deterministic causes (1 *b*)

$$\begin{aligned} A &\rightarrow a \\ A' &\rightarrow a' \\ A'' &\rightarrow a'' \end{aligned} \tag{1b}$$

without appealing again to the very thing to be explained, namely to a statistical distribution ratio?

(2) Or are such hopes unreasonable and futile?

The terms 'reason', 'hope', 'explain' point again to cognitive inquiry and personal interpretation. As to possible modes of interpretation one may distinguish between *hard* and *soft determinists*. All determinists agree with Einstein when he said: 'I cannot believe that the Lord plays dice with the world.' If he had added: 'nor that He has ever played dice', it would have been a clear and consistent statement of *hard determinism*. But how would a hard determinist account for the strange (from the deterministic viewpoint) fact that in all cases of statistical distribution, of gas molecules, of clicks in Geiger counters, of automobile accidents, of raindrops falling per square yard, etc., one always finds *agreement between statistical fact and mathematical theory of random*, as to average distribution as well as to fluctuation? Does this not indicate that at least *once* there must have occurred a *mixing* of (initial) conditions in which random was supreme? This is what *soft determinists* concede with their 'hypothesis of molecular disorder'. But if they admit a mixing of individual conditions at *one* remote instant long ago, passed on deterministically through the ages, then it is hard to see a world-shaking step from there to the concession that such mixing may take place at other times, too, namely, whenever a new statistical situation in agreement with random law occurs, in gambling casinos, in 'classical' molecular distribution, in atomic quantum processes such as radioactive disintegration. Niels Bohr(6) holds that the uncertainty of the outcome in an individual atomic process dominated by the constant h is something *essentially* different from the uncertainty in classical kinetic gas theory where it is due *only* to the complexity of the situation. Yet the difference is in the thinking about the facts, not in the facts. What essential difference is there between the random distribution of α-particles escaping from a radium preparation, and that of small balls escaping from a large box with a small hole? In both cases one observes an exponentially decreasing rate of escape, and the same fluctuations. Planck's constant h is certainly of quantitative importance for the magnitude of the 'escape hole' in a radium atom; but it is not essential for the statistical dispersion as such. Likewise, the alleged distinction that 'in ordinary games of chance we could, if we would, reduce chance distribution to deterministic causes, whereas in quantum games we could not even if we would' is illusory, as the following analysis may show.

2. *An ordinary game of chance*

The game may be the dropping of balls from a chute upon the edge of a knife. If the chute is aimed at an angle A clearly to the right of the edge, all balls will drop to the right of the knife. The same holds for the left. Experience shows, however, that there is always a small but *finite* range ΔA of *physical* (not ideal!) aim inside of which not all balls drop to the same side. Instead, for every angle A of physical aim (obtained, for example, by a micrometer screw) inside the small range ΔA, one observes a certain statistical frequency ratio between r- and l-balls. This ratio $r{:}l$ varies from 100:0 to 0:100 when the physical angle of aim (voluntary preparation of the screw) is varied from one to the other end of the range ΔA. It thus appears as though one and the same physical cause A, if it is within the range ΔA, may produce two different effects, r or l respectively, individually at random, yet controlled by a statistical ratio which depends on the geometrical position of A within ΔA.

In spite of the familiar nature of this game of chance, a

determinist ought to be greatly disturbed by the discovery of the *conformity* between statistical distribution on the one hand, and the mathematical theory of random events on the other, a conformity which pertains not only to average frequency ratios (such as a 30:70 frequency ratio of $r{:}l$ when the screw is turned 30 per cent over the range ΔA), but also for the fluctuations away from the average in agreement with the mathematical error law of Gauss. This apparent 'pre-established harmony' between fact and theory, this *statistical co-operation* of otherwise independent events, is a most serious challenge to the deterministic doctrine.

The usual comment upon the ball-knife game runs as follows: Even within the uncertainty range ΔA of physical aim (turn of the screw) each *r*- and *l*-event has its distinct cause, be it a slight deviation of aim, a minute dislocation of the blade, or a perturbation of the ball during its flight; and each cause in its turn is the effect of an earlier cause. That is, an individual *r*-event is but the terminal member of a deterministic chain . . . $rr\underline{r}$ reaching into the infinite past. Each final event could have been predicted by a superior spirit with insight in all the causes. Similarly, the date of an individual mortality could be foreseen by an ideal doctor, and an automobile accident is not an 'accident' but has its definite causes.

All this may be so, but it avoids the heart of the problem which is that of the *conformity between a priori mathematical theory and statistical fact*. It shifts responsibility for today's conformity to that of yesterday. Let us not forget that 'co-operation' of individual events in a statistical ensemble is a most unexpected fact. Its discovery by our ancestors marked an important step forward in natural philosophy, from the creed in a hostile and whimsical world, towards a *cosmos* in which even the god of thieves and gamblers is bound to obey rules of fair play. And what is fair play in a game? Conformity between average numerical ratios and geometrical ratios! Fair dice have equal sides, hence claim equal chances. And equal sections of a wheat field ought to receive equal numbers of raindrops. But although conformity of geometrical and numerical ratios is plausible, it is not a *causal* explanation of that conformity. It is not even correct, it cannot be correct for small numbers, for example, for only seven throws of a die. Why, then, is it correct for large numbers? There might be an individual cause for this ball ending up on the right and another ball on the left of the edge, for example a corresponding small air current. But what causes about fifty right currents of air to every fifty left ones? Answering that the air currents are distributed 'at random' is not a causal explanation but simply a statement of fact, taking the *explanandum* for granted.

The only causal explanation for the harmony between mathematical expectation and statistical fact, in short for 'statistical co-operation' of independent individual events, would be the following:([7]) Once upon a time there was a *demon*, representing Max Black's free-acting agent, who in the ball-knife game first started two causal *r*-chains, then one *l*-chain, then three *r*-chains. Thereupon, realizing that he had given too much preference to *r*-chains, he started four *l*-chains in a row, deliberately causing a pseudo-random distribution, in order to lure present-day observers away from the true deterministic faith and to deceive scientists into accepting the heresy that 'the Lord plays dice with the world', or at least has played dice once upon a time. To help his evil work, our demon may have used the decimals of the number π, with even decimals for *r* and odd ones for *l* or Karl Popper's([13]) procedure, to give his plan an even less intentional look. Only when we are prepared to accept a *deus ex machina* as the cause for a distribution which only *looks* like random, can we pretend to have saved the deterministic ideology in spite of random appearances. But why such a demon should be admitted only in the theory of 'ordinary' statistical series and insurance, whereas in case of quantum series 'real true random' must be accepted, is beyond my capacity to grasp.

Therefore I cannot see how and why quantum theory has

introduced an essentially *novel* situation with deep philosophical implications. Let us better concede what philosophers of science have seen long ago, that there is *no causal explanation* for the harmony between statistical dispersion and mathematical random theory, hence that 'statistical co-operation' must be accepted as a basic feature of the world in which we live. The usual talk about concealed causes is but an evasion; it merely blames present statistical co-operation on past co-operation in an infinite regress. Hence I deny that 'in ordinary games we *do not* know and in quantum games we *can not* know the outcome'. The *cannot* holds in both cases, all classical thinking to the contrary.

3. *The Quantum Game*

We now discuss a slight modification of the 'ordinary' macroscopic game of chance described before. Instead of balls aimed at a knife edge, consider a certain 'kind' of 'atoms' aimed at a 'filter' which can either 'pass' or 'reject' it, or in some other way either react with *yes* or *no*, depending on the 'state' of the atom. The terms in quotation marks are to be understood in the most general sense.

Among the class of all 'states' S which our atom is capable to possess we may distinguish the subclass A of states which are passed, and the subclass *non-A*, written $\bar{A}$, of states which are repelled by this particular filter. The latter may then be denoted as an A-passing filter, or shortly as an A-filter.

To the remark that the 'filter' is here defined with reference to 'states', and 'states' with reference to 'filters', I have this to say: A theoretical structure works with mathematical and verbal symbols without specific content. Thereafter it may be associated with a 'model', a physical counterpart. The structure of 'states' developed here is exemplified in quantum physics by 'states of position at a given time', by 'states of momentum', and so forth, but certainly not by states of health or states of knowledge of some observer. The formal structure may be built without always referring to familiar experiments—just as the geometer applies the concept of points and straight lines without always giving an operational definition of these terms.

By separating the subclass A of states from the subclass $\bar{A}$, the filter characterizes A and $\bar{A}$ as discernible, as entirely different, written $\bar{A} \neq A$.

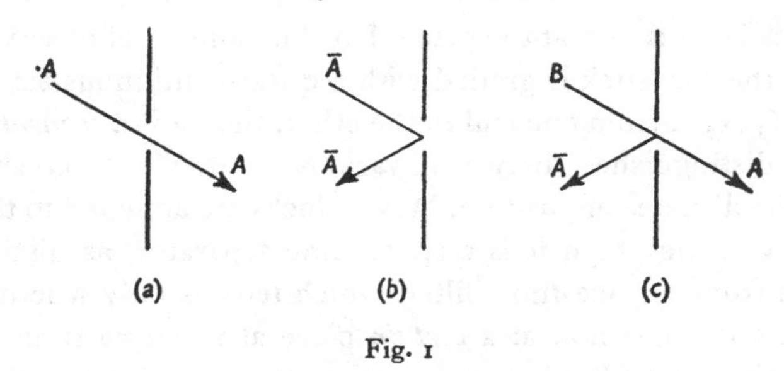

Fig. 1

Experience shows, however, that there is a third subclass of states intermediate between A (always passed) and $\bar{A}$ (never passed), namely states B which are sometimes passed, sometimes rejected by the A-filter, as depicted in Fig. 1*c*. States B showing this reaction can neither be regarded as 'equal' to A nor as 'entirely unequal', that is separable from A. We shall denote such states B as 'fractionally equal' to A, written $B \sim A$. A *fractional equality degree* between A and B can be defined quantitatively as the statistical *passing fraction* of B-state atoms through the A-filter. However, since the concept of equality is mutual, this definition makes sense only when the B-state atoms pass through an A-state filter at the same fractional ratio as the A-state atoms pass through a B-filter. This *symmetry* requirement is basic for the probability metric of Chapter III. It will be seen in section 10 that it is the statistical correspondent of the *reversibility* of processes in classical mechanics; therefore it can be regarded as plausible and 'natural'. The statistically ruled phenomena of Fig. 1*c* are known to the quantum theorist as the *splitting effect*.

The question has been asked: 'what happens when an atom gets stuck in the filter?' To which I reply: 'getting stuck is the same as not passing.' Remember that the term 'filter', as illustrated in Fig. 1, is meant only in a generalized sense; it refers to any instrument constructed for the purpose of distinguishing, that is separating a state A from a state non-A. For example, a yardstick with a mark at a place A is a position *filter* in so far as it tells us whether a particle roaming along the yardstick is, or is not, at the place A at the moment of observation. If the yardstick is graded with a quasi-continuous set of marks $A_1\ A_2$ . . from one end to the other, then it is a *separator* which distinguishes between various mutually exclusive 'orthogonal' states of position. When clocks are attached to the graded yardstick then it is a space-time separator, as distinguished from a space-time 'filter' which tells us only whether a particle is, or is not, at a *certain* place at a *certain* time. A Nicol prism is a *filter* for one definite state of polarization of photons. A doubly refracting crystal is a *separator*, for two mutually exclusive (= orthogonal) states of polarization. These examples may suffice to answer objections about 'stuck' particles. Another question has been: 'do not your filters violate the uncertainty principle?' On the contrary, 'my' filters offer the simplest general example of uncertainty of which the Heisenberg qp-uncertainty is but a special case.

The Stern-Gerlach separator (Fig. 2) splits an incident beam of silver atoms in two separate beams with parallel and anti-parallel orientations of the atoms with respect to a magnetic field. When it is maintained here that this 'splitting effect' is essentially of the *same nature* as the ordinary ball-knife game, quantum theorists assume an air of superiority and insist that the splitting of an atomic beam into separate components is 'something entirely different', being primarily a decomposition of an original ψ-wave into 'polarized components' with definite phases, and with the observed intensities of the components being *merely* the absolute squares of the ψ-amplitudes. The writer agrees with the mathematical part of this consideration, the more so as he had been the first to point out the perfect analogy of the magnetic splitting effect with optical polarization.(8) And yet he must object to the opinion that there is an essential difference between a series of ordinary balls separated into r- and l-components by a knife, and the magnetic splitting of an atomic beam. The ψ-phases in the latter case are quite irrelevant for the splitting of an incident atomic beam by *one* field; they are significant only for the metric *relation between several sets of probabilities*, for example, those occurring when a beam is sent through *several* magnetic fields of different orientation (refer to the general probability metric of Chapter III). But ψ-phases are not possessed or carried along by individual atoms and the quantum feature that the uncertainty of position increases when that of momentum is narrowed down—important as it is for mechanics—is a technicality within the general experience that one and the same originally prepared state A may be followed by different states a, or a', or a'' in classical as well as in quantum games.

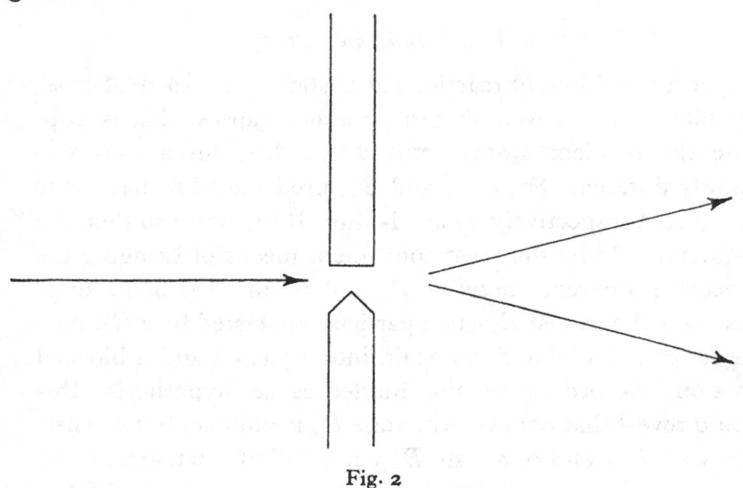

Fig. 2

The idea of *filters* in the most general sense is helpful for a simple and unsophisticated approach to the quantum theory. The 'filter' of Fig. 1 is a generalization of the semipermeable diaphragm known from the thermodynamic theory of gas mixtures. For this reason the writer once thought that quantum theory might best be deduced from a thermodynamic basis. This is possible indeed (refer to section 14). Nevertheless, following Gibbs and Boltzmann, mechanics is primary, and thermodynamics is but a consequence based on the additional hypothesis of 'molecular disorder'. In the deduction of quantum theory presented here I therefore have relegated thermodynamic considerations to a place of secondary rank, whereas in an earlier presentation(9) the postulate of entropy continuity was placed at the beginning.

It ought to be emphasized that thermodynamics deals with equilibrium or quasi-equilibrium states, and with quasi-static processes controlled by semi-permeable diaphragms. In contrast, our considerations also admit non-conservative states, for example positional states at a certain time, etc., so that our 'filters' are generalizations of the diaphragms of thermodynamics.

4. *Von Neumann's Proof*

Suppose the different reactions of B-state atoms to an A-passing filter, Fig. 1*c*, were due to concealed causes; that is, suppose the incident atoms actually are distributed over two slightly different states, B_A and $B_{\bar{A}}$, predestined to pass or to be rejected respectively by an A-filter. If this were so then the A-filter would be the most convenient means of bringing the concealed difference between B_A and $B_{\bar{A}}$ into the open. Suppose now the passed B_A-state particles are tested by a C-filter; the latter will divide them again into a passed and a blocked fraction. According to the hidden-cause hypothesis, this would reveal that not even the state B_A is uniform but consists of a state B_{AC} and of a state $B_{A\bar{C}}$ at a definite statistical ratio, the former being passed and the latter rejected by a C-filter. Continuing in this fashion it becomes obvious that the concealed cause hypothesis would ask us to believe *first* that every single atom is predestined for a definite reaction in any possible future filter test, and *second* that the fates of various atoms are co-ordinated in advance so as to yield definite statistical ratios plus Gaussian fluctuations of passed and blocked numbers, all this in harmony with mathematical error theory. The *first* is indeed what determinism wants us to believe; but the *second* would require a foresight of really incredible dimensions, or a gigantic fraud staged by an evil demon in order to feign random *appearance* in every single test. It would be just the opposite of what adherents of determinism try to defend with the statement: 'the world simply is not in an exception state of order.' In view of the observed 'statistical co-operation', that is of the harmony between statistical fact and mathematical random theory, determinism cannot be maintained as a physical theory in spite of 'its protean capacity to elude refutation', by appealing to ever more concealed causes.

Since the considerations above apply to a succession of 'quantum tests' of atomic particles, as well as to a succession of ball or dice games, I cannot see the need for an extra *proof* of the irreducible statistical character of quantum events in atomic tests. When von Neumann([10]) proposed such a proof many years ago, the situation was entirely different from the present one. At that time the success of the Schrödinger wave theory seemed to suggest the unitary model of matter waves evolving deterministically under a time-dependent differential equation, with wave crests playing the role of particles. When Born proposed the likewise unitary statistical particle interpretation, a formal proof of the legitimacy of this statistical interpretation seemed to be desirable. It was overlooked, however, that a given formalism always allows a variety of interpretations, unless additional restrictions are introduced. And von Neumann's 'proof' that the quantum formalism requires a statistical interpretation rests on statistical assumptions, as shown by de Broglie([11]) and Feyerabend([12]); that is, his proof is circular, hence useless. *Vice Versa* if one hopes that this recent refutation of von Neumann's proof reopens the path towards a return to determinism in quantum physics, he may as well abandon this hope. For statistically distributed events, whether observed in ordinary or in quantum experiments, can *never* be deterministically explained, unless one resorts to the hypothesis of a *deus ex machina* who has produced the *appearance* of random. Therefore Charles S. Peirce, in his essay, *The Doctrine of Necessity Examined*, of 1892, replied to a (soft) determinist: 'You think all the arbitrary specifications of the universe were introduced in one dose in the beginning, if there was a beginning. But I, for my part, think that the diversification, the specification has been continually taken place.' For a recent evaluation of the internal contradictions of the deterministic doctrine refer to K. R. Popper([13]), 'Indeterminism in Quantum Physics and in Classical Physics'. See also 'Determinism and Freedom in the Age of Modern Science, a Philosophical Symposium', New York, 1958, quotation 2 above.

When the 'miracle of co-operation' of individual events in statistical ensembles is once accepted as irreducible (demons being excluded), then there is no reason for amazement every time that another example of such co-operation is found. For instance, the famous experiment of matter diffraction through two slits (or through a crystal) shows intensity maxima and minima obviously built up by individual arrivals of particles one after the other. Here it has been asked how one particle can 'know' what other particles have done, or will do, at other places and at other times so as to produce the diffraction pattern in a statistical manner. The permanent (not statistical) co-operation of the two slits is a quite different matter discussed in section 27.

5. *Reproducibility of a Test Result*

When among 100 incident *B*-state particles seventy pass and thirty are rejected by an *A*-filter (Fig. 1*c*) then we have a right to ask: *how* do the seventy *B*-state particles manage to pass a filter which is built so as to pass *A*-state particles with 100 per cent certainty? There is a simple answer to this question: those seventy *B*-state particles *jump*, in contact with the *A*-filter, to the new state *A* and thereby become entitled to pass; similarly, the thirty rejected *B*-state particles jump to a state *non-A* and thereby forfeit their right of passage. The newly acquired states are indicated on the passed and the rejected beam in Fig. 1*c* as *A* and $\bar{A}$. After their reaction the particles remain in their newly acquired states, unless they are challenged again by another filter, C, where they have to chose between jumping to C or $\bar{C}$.

Those jumps from B to A and $\bar{A}$ at a certain statistical ratio present us with a startling 'quantum phenomenon' which calls for an explanation, that is for a reduction to a more elementary and plausible general basis. The desired explanation is furnished by the principle or postulate of *reproducibility of a test result*: 'An object which has once responded with *yes* (or *no*) to a certain testing instrument will respond *yes* (or *no*) again when the test is repeated," unless another test (or a time interval, refer to section 11) is interposed. In the present case: 'B-state particles once passed (once rejected) by an A-filter will be passed (be rejected) by another A-filter again with certainty.' *Hence*—this is the conclusion—they must have acquired the new state A (or $\bar{A}$) during their first passage (or rejection) and then *must stay in the newly acquired state.* Similarly, a photon which has once passed a Nicol prism will pass another parallel Nicol prism with certainty; hence it must have acquired a state of polarization parallel to the two Nicols in the act of passing the first. Reproducibility of a test result may fail for position tests within dimensions of 10^{-13} centimetres and thus necessitate a revision of present-day quantum theory for nuclear phenomena.

6. *One or Many Games of Chance?*

Classical *statistical* mechanics is based on the tacit assumption that once upon a time a random-like state had been established (by a demon?), and that only from there on all events were determined. Quantum theory maintains that every test involves a new independent random distribution. The difference of the two standpoints may be illustrated by the following example.

Imagine a great number, N, of arrows distributed as radii of a circle in a random fashion. Suppose this 'star' of arrows is viewed through a circular glass A divided into three equal sectors A_1 A_2 A_3. One can 'predict' here that each sector will cover $N/3$ arrows. Accordingly, the betting odds of finding an individual arrow within section A_1 or A_2 or A_3 amount to 1/3 each. The statistical predictions in this and similar games are based on the initial random distribution of the arrows over all directions; and these predictions may be confirmed simply by *viewing* the arrows, without thereby affecting the state of any arrow so viewed. This is the classical situation.

When the N macroscopic arrows are replaced by micro-arrows representing the spin directions of N electrons, then not even a demon can distribute them over all directions of a circle simultaneously. Spin arrows can be laid out only by passing them through a linear 'slit' or 'filter', for example a magnetic field held in a certain direction, from which the spins emerge in two opposite directions, parallel and anti-parallel to the field, according to chance, anew in each experiment, without *deterministic* appeal to previous distributions being possible. When the magnetic field is held in the N.–S. direction, the electrons line up their spins in the N. and S. direction only. When they are later tested by a second field, this time of NE.–SW. direction, each spin is compelled to *jump* from its original N. or S. direction either to the NE. or to the SW. direction. For an individual arrow it is a game of chance whether it will jump to the NE. or to the SW. direction; only the average frequencies of the various jumps have definite magnitudes. (They are equal to the cosine squares of half the angle between the original and the final direction.) The macroscopic testing instrument, in the present instance the magnetic field, plays an active part in compelling the tested particles to jump from their original direction to one of the two directions peculiar to the magnetic field. Which of the two directions is reached by an individual particle is a matter of probability. Similarly, when a position meter has located a particle at a certain co-ordinate point, x, or within a small range Δx, and when this observation is followed by a test of the x-momentum of the same particle, then it is a matter of probability which momentum value will actually turn up in such a test.

And no experimental arrangement can put particles into a state in which both position and momentum have definite values simultaneously in a reproducible fashion with certainty (Heisenberg). That is to say, position and momentum are mutually *incompatible* observables. Similarly, spin orientation along the N.–S. direction and spin orientation along the NE.–SW. direction are mutally incompatible observables.

Big arrows can be distributed at random over an infinite number of directions; spin arrows can be distributed over two opposite directions only. In spite of this difference, however, I cannot agree with Niels Bohr[14] when he writes in his *Discussion with Einstein* (p. 203 l.c.): 'It is most important to realize that the recourse to probability laws under such (atomic) circumstances is essentially different in aim from the familiar application of statistical considerations as practical means of accounting for the properties of mechanical systems of great complexity. In fact, in quantum theory we are presented not with intricacies of this kind, but with the inability of the classical frame of concepts to comprise the peculiar feature of indivisibility or "individuality", characterizing the elementary processes.' In my view it is on the contrary most important to realize that there is *no essential* difference between the statistical phenomena observed in plain 'ordinary' games and in atomic tests. There is no *essential* difference between the right-or-left ball game of section 2, and the pass-or-repel atomic game of section 3, nor between the escape of balls from a box with a hole, and the escape of α-particles from radium atoms with a 'hole'. It is delusion to think that in case of ordinary games 'we could if we would' restore determinism, whereas only in quantum games 'we could not, even if we would'. When P. Bridgman[15] remarks that 'the seeds and sources of the ineptness of our thinking in the microscopic range are already contained in our present thinking in the large-scale region and should have been capable of discovery by sufficiently acute analysis of our ordinary commonsense thinking', I agree: the idea of getting along with determinism in ordinary games shows the ineptness of our thinking indeed. 'Statistical co-operation' of individual events in a statistical ensemble, whether ordinary or quantal, does not admit of a deterministic explanation. Therefore I cannot follow the usual view that there is an *essential* difference between ordinary and quantum games of chance.

7. *Indeterminacy, a consequence of continuity*

The realization that the deterministic doctrine fails to account for the *harmony* between statistical fact and mathematical error theory based on geometrical symmetry may be very disturbing to our ingrained ways of thinking about a causal world. However, 'a conceptual scheme is either modified or replaced by a better one, never abandoned with nothing taking its place' (James Conant). It is fortunate, therefore, that the shock of having to accept indeterminacy on the evidence of any game of chance, can be cushioned by another principle, familiar to us from childhood days, which pleads in favour of accepting, and even compels us to accept, statistical law irreducible to individual determinism. I refer to Leibnitz's principle of *cause-effect continuity* which in his own words reads: 'when the cases approach each other continuously and finally get lost in one another, then the events in the sequel do so also', or in modern language: 'A finite change of effect requires a finite change of cause.' As will be seen presently, continuity of cause and effect is incompatible with unrestricted determinism.

Consider our ball-knife game again. It would be 'unphysical' to expect a change, from all balls dropping to the right to all balls dropping to the left, taking place suddenly at one single angle of aim, when the aim is swept gradually from right to left. Abrupt changes of effect may be imagined theoretically, but they do not occur in nature. Rather, when *physical* balls are aimed at a *physical* knife-edge by means of a *physical* mechanism, then, for the sake of continuity there ought to be a *finite* range ΔA of the aiming angle A so that inside of ΔA a gradual change of the $r{:}l$ ratio from 100:0 to 0:100 takes place via intermediate ratios of $r:l$. The only conceivable way to produce intermediate ratios, however, is the *statistical* way.

According to a similar consideration one will have to *expect* those phenomena of atomic physics which in section 3 were described as the *splitting effect*. Indeed, suppose particles of a certain 'kind' are in a 'state' A so as to pass an A-filter. Suppose, further, the particles can also be brought into another state so that they now will be unable to pass the A-filter; we denote this other state *non-A*, written $\bar{A}$. The continuity principle now asks us to expect that 'between' the states A and $\bar{A}$ there ought to be other states B so that B-state particles are neither all passed nor all blocked by the A-filter, but sometimes passed and sometimes blocked. Now, if the passed: blocked ratio is not to be quite erratic, it can only be a definite *statistical* ratio. And again, according to the continuity postulate, there ought to be a continuous series of B-states with a continuous series of ratios between passed and blocked B-particles, ranging from 100:0 defining $B=A$, to 0:100 for $B \neq A$, with intermediate cases of *fractional equalities*, written $B \sim A$. The statistical *passing fraction* may then be taken as a quantitative definition of the 'fractional equality degree' between A and B. These statistical effects, pictured in Fig. 1*c*, are usually taken as matters of fact; they can be seen as consequences of the continuity principle which bridges the sharp contrast between equality and inequality, by forcing us to accept the important concept of a *fractional equality* $A \sim B$ between two 'states' of an object. Fractional equality ought to be mutual; this request is satisfied by the *symmetry* discussed in section 3.

Generally speaking, *between* the reaction 'yes' or passing, and the reaction 'no' or blocked, there can only be the reaction 'undetermined'. But instead of resorting with Reichenbach([16]) to a three-valued logic, ordinary logic can be retained and be applied to statistical ratios, from 'all yes' to 'all no', and with indeterminacy in the intermediate cases. *Causal anomalies are not logical anomalies*. On the contrary, the occurrence of acausal events controlled by statistical law is foreshadowed by the postulate of cause-effect continuity. And the apparent paradox that the *continuity* postulate together with that of reproducibility requires *discontinuous* jumps from state to state (Fig. 1*c*) is removed by the realization that those discontinuous transitions from state to state in tests are controlled by chance rather than by cause-effect relations. We thus have cause-effect continuity together with acausal discontinuity.

In order to understand quantum theory it is much better to begin with an analysis of such simple experiments as the splitting effect of Fig. 1*c* and Fig. 2, rather than starting from Planck's $E=h\nu$ which is a remote consequence of several elementary postulates in combination. A simple straightforward avenue to the intricacies of quantum mechanics has been obstructed, however, by the historical development which derived its impetus from spectacular discrepancies between microphysical fact and classical expectation, rather than from a critical analysis of the ineptitude of classical ideas already in simple 'ordinary' games of chance. With the sole exception of Bohr's correspondence principle, the emphasis has always been on the dramatic and revolutionary; and it had to be so in order to justify the break with the Cartesian tradition. Then at long last it all seemed to condense in the Schrödinger equation, today placed at the entrance gate to quantum theory. But this is like beginning a study of mechanics with $E=mc^2$ rather than with falling apples and swinging chandeliers, or electromagnetism with the Maxwell field equations rather than with Coulomb's law. Intricate developments involving complex imaginary ψ-functions, periodic in co-ordinates and momenta according to $p=\mathrm{h}/\lambda$ and $E=\mathrm{h}\,\nu$, ought to be placed at the end, not at the beginning of the theoretical exposition. I therefore have reversed the usual line of approach. Our first concern is to develop the general schema of probabilities, its 'why' (Chapter II) and 'how' (Chapter III). Special applications to conjugate observables q and p are left to Chapter IV, where the wave-like h-dominated features of quantum dynamics are developed.

8. *The axiomatic method*

The axiomatic method in *mathematics* starts out from an abstract schema of logical reasoning, stripped of all specific meaning, a symbolism which can be handled as easily as a set of equations. Euclid's and Hilbert's axiomatization of geometry are shining examples. Another example is Cantor's theory of infinite classes where antinomies arise warning us that the system of axioms ought to be free of inherent contradictions, yet be exhaustive enough so as to cover a given field. In mathematics it is the formalism of the symbols which takes precedence over applications to subject matter. One and the same formalism may be applied to a variety of models, an example being projective geometry where the symbols for 'points' and 'straight lines' can be interchanged.

The axiomatic method in *physics* proceeds by a different road towards a different goal. It starts from general principles of an empirical character condensed from broad experience: 'nature is such that always . . .'. By combination of several general principles and their application to various subject matter, theoretical physics arrives at more special and involved laws. The physical laws are often expressed in the sign language of mathematics so as to yield remote conclusions by schematic calculation without definite physical significance attached to each step of the mathematical procedure. If experience should not confirm the theoretical predictions, some of the underlying general postulates must be sacrificed or modified—without thereby lending eternal validity to those retained.

A standard application of the axiomatic method in physics is the science of thermodynamics, large parts of which are derivable from the first and second law, that is from the axioms that energy and entropy are 'state variables' and are functions of other state variables such as pressure, volume, etc., whereby energy is constant and entropy is increasing in closed systems. But since the second law itself is somewhat puzzling, various efforts have been made to prove its validity on the basis of simpler and more immediately acceptable principles, for example that heat never flows from a cold to a hot place, or that a perpetual-motion engine transforming heat without loss into work cannot be constructed. By far the simplest derivation of the second law has been given by Caratheodory on the basis of his axiom: 'in the neighbourhood of every state *A* of a body there are other states *B* which cannot be reached from *A* in an adiabatic fashion.' In spite of its great simplicity, Caratheodory's axiom and proof was coolly received as being too far removed from direct experience. Today, forty years later, most textbooks have adopted Caratheodory's approach as the shortest and most convincing way to obtain the second law of thermodynamics. The goal of the present study lies in the same direction of deducing quantum mechanics from general propositions.

9. *Summary*

In contrast to the purely academic Laplacian doctrine of universal determinism, the empirical law of causality maintains that, when an object has been repeatedly *prepared* in the same manner as closely as possible, then it will always produce the same impression on another body likewise equally prepared as close as possible, serving as a testing instrument. In short, equal causes, equal effects. This empirical law, with emphasis on 'as closely as possible', is challenged by the experience in games of chance, where equal preparation leads to different results. The observed *harmony* between random sequences dominated by fixed statistical averages on the one hand, and the mathematical *a priori* expectation based on geometrical or other symmetries of arrangement on the other hand, in short, the *statistical co-operation* of independent individual events yielding definite averages and fluctuations away from the average (Gauss error law) is a miracle from the deterministic point of view. Causality can at best ascribe present to past statistical co-operation in an infinite regression. There is but one way to account for random distribution of individual events in a statistical sequence as not being 'true random', namely the assumption that a demon has deliberately (= causally) produced a pseudo-random distribution in harmony with mathematical chance law 'at the beginning of time'. Since a *deus ex machina* is no substitute for a reasonable theory, statistical co-operation in harmony with which mathematical *a priori* expectation must be accepted as basic and irreducible, together with the notion that random events take place continually, as C. S. Peirce remarked long before the quantum age. There are not sufficient causes for every event.

On the other hand, there is sufficient reason for the lack of sufficient causes when one accepts Leibnitz's principle of *cause-effect continuity*. This principle leads first to the conclusion that the reaction 'always yes' or 'always no' respectively should be bridged by intermediate reactions 'sometimes yes and sometimes no', that is, by cases of *indeterminacy* in an individual test, dominated by statistical averages. When applied to 'states' tested by their reaction to 'filters' in the most general sense, one arrives at the concept of a *fractional equality* between two states A and B, written $A \sim B$, intermediate between the limiting cases $A = B$ and $A \neq B$, that is, between inseparability and separability of A and B. The fractional equality is quantitatively definable as the *passing fraction* of A-state particles through a B-passing filter, and vice versa, assuming the empirically confirmed principle of *symmetry*. With the additional postulate of *reproducibility* of a test result, the passing fraction $P(A \rightarrow B) = P(B \rightarrow A)$ becomes the *probability of transition* from state A to B in a B-test, and from B to A in an A-test. All this is exemplified by the splitting effect as a fundamental quantum phenomenon without involving Planck's quantum h.

CHAPTER II

STATES, OBSERVABLES, PROBABILITIES

'Agreement with experimental facts must not be sought in the initial steps of theoretical analysis but in the final results.'

EINSTEIN

10. *States, magic square tables*

The previous chapter dealt only with the relation between a single pair of states, A and B, in the three cases of equality, inequality, and fractional equality. The same considerations shall now be applied to the class of all states of an object. (An operational definition of the term 'state' will be given in section 11.) Imagine that all 'states' of a given object (atom) are listed one by one on separate cards. Order may be brought into this chaos by the following procedure.(1) Draw one card and denote the state it represents as A_1. Draw a second card representing a state A_2, selected so as to be totally unlike, that is separable, from A_1. The third state A_3 shall be chosen so as to be totally unlike both A_1 and A_2. Continuing in this fashion one may collect a set of mutually unlike, that is entirely separable states A_1 A_2 A_3 . . . When after a number of draws no state can be found any more which is entirely unlike *all* of the previously drawn states, then one has collected a *complete set* of mutually unlike, or separable, or 'orthogonal' states satisfying the conditions

$$P(A_k, A_{k'}) = \delta_{kk'} \begin{matrix} = 1 \text{ for } k = k' \\ = 0 \text{ for } k \neq k' \end{matrix} \qquad (1)$$

where P indicates the fractional equality degree (section 3).

The next draw will yield a state, denoted as B_1, which is neither identical with any of the former states A, nor entirely different from *all* of them; otherwise it would have been included in the set A. The state B_1 must therefore be *frac-*

tionally equal to some, if not to all of the states A. The same holds for the next state B_2 chosen so as to be entirely unlike B_1. Continuing in this fashion one arrives at a complete set of mutually orthogonal states B_1 B_2 . . , comprising the set B. Thereafter one may collect other orthogonal sets of states C, and D, and so forth, until all states are arranged in orthogonal sets. The division of all states into orthogonal sets is not unambiguous, a fact connected with 'degeneracy', but irrelevant in the present connection. The mutual equality fractions P between the members of each single set satisfy orthogonality conditions similar to (1), whereas the mutual equality fractions between any two states belonging to different sets are fractions of unity, $0 \leqslant P < 1$. As seen in section 3, they signify transition probabilities in tests; they are compiled in tables or matrices:

$$(P_{AB}) = \begin{Bmatrix} P(A_1, B_1) & P(A_1, B_2) \ldots \\ P(A_2, B_1) & P(A_2, B_2) \ldots \\ \ldots & \ldots \end{Bmatrix} \qquad (2)$$

In the same manner one may compile tables (P_{AC}), (P_{BC}), and so forth.

Interpreted physically, the orthogonality condition (1) implies that an A_k-passing filter blocks particles in all other states $A_{k'}$ of the same set A. On the other hand, the A_k-filter passes a certain percentage of incident B_j particles, by throwing them from state B_j to A_k, with probability $P(B_j, A_k)$. In practice one does not depend on selective 'filters'. There are instruments, so-called *separators*, which distribute incident particles from their common original state A_k directly over the states B_1 B_2 B_3 . . , as indicated in the schematic Fig. 3, and exemplified by the Stern-Gerlach 'splitting effect'. The sum of the probabilities (= relative statistical frequencies) of jumping from one state A_k to the various states B_j of the orthogonal set B is *unity* as a matter of course:

$$\Sigma_j P(A_k, B_j) = 1 \qquad \text{for every } k = 1, 2, 3, \ldots \qquad (3a)$$

That is, the sum of the P's in any one *row* of the matrix (P_{AB}) is unity.

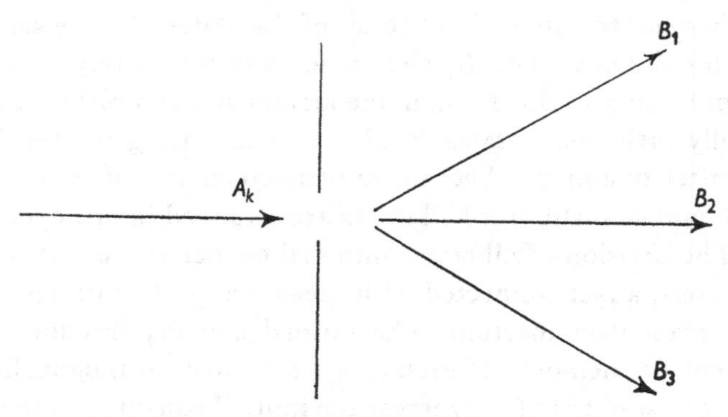

Fig. 3. *B*-separator.

There is, however, another far from self-evident rule, discovered in atomic experiments, namely the rule that the sum of the elements in every *column* of the matrix (P_{AB}) is also unity:

$$\Sigma_k P(A_k, B_j) = 1 \qquad \text{for every } j = 1, 2, 3, \ldots \qquad (3b)$$

That is, the probabilities of a system *arriving* in one state B_j from various states A_k of a complete set A add up to unity. This strange-looking rule follows immediately from the elementary theorem of a two-way symmetry for the transition probabilities:

$$P(A_k, B_j) = P(B_j, A_k) \qquad (4)$$

Indeed, according to this rule, the columns of the table (P_{AB}) are the rows of the table (P_{BA}). The two-way *symmetry relation* (4) itself must be introduced into the theory as a separate postulate. It can be justified, however, by its *correspondence* to classical mechanics, as being the statistical counterpart of the *reversibility* of classical mechanical processes. What has been said above of the matrix (P_{AB}) of course holds for any P-matrix, such as (P_{BC}).

The unit sum rule for the rows and columns of the P-tables leads to the further conclusion that all orthogonal sets of states of a given system (atom) have *common multiplicity* M; that is, each set consists of M members, so that all P-tables are quadratic with M rows and M columns. The proof runs as follows:

Suppose the set A has M members $A_1\, A_2 \,.\,.\, A_M$, and the set B has N members $B_1\, B_2 \,.\,.\, B_N$. If one sums up all members of the table (P_{AB}) row by row, its M rows sum up to M. If the summation is carried out column by column, the N columns sum up to N. Hence, $M = N$, that is *the P-tables are quadratic.*

Because of the sum *unity* for every row and every column, the P-tables may be denoted as *unit magic squares.* Equations (1) to (4) represent the *formal framework* on which the general metric structure of probabilities (Chapter III) and the quantum dynamics (Chapter IV) are erected. The rest of the present chapter is devoted to physical applications.

11. *States, observables, particle kinds*

Suppose an atom, originally ascertained in the state A_1, emerges from a B-separator test in the state B_3, having jumped from A_1 to B_3. In this new state it has lost all remembrance of its former state A_1; indeed, if one should subject it to an A-test again, it may emerge in any of the states $A_1\, A_2 \ldots,$ with probabilities $P(B_3, A_1)$, $P(B_3, A_2)$, etc. Hence it is impossible to produce physical conditions in which the two states A_k and B_j are realized simultaneously in a *reproducible* fashion. For this reason, the two sets of states A and B are denoted as *mutually incompatible.* On the other hand, when a state A_k has been ascertained in an A-test, and the particle is subjected to a second A-test, then it will be found in the same state A_k again. We denote this as the *reproducibility of a test result* concerning *one* state.

The Stern-Gerlach magnetic field separator (Fig. 3) may serve as an illustration. Atoms entering the instrument in any original state of orientation are forced by the magnetic field to choose among several discrete orientations $B_1\, B_2 \ldots$ with respect to the direction B of the magnetic field; they may (in theory, if not in practice) be collected separately in these newly acquired states of orientation. The Stern-Gerlach *separator* can be remade into a 'filter' for B_3-state particles by screening off the other B-components. Suppose the B_3-state particles are sent into a second magnetic field separator of field direction C enclosing an angle with the field direction B of the first separator. The B_3-particles now have to jump to states of orientation C_1 C_2 . . . with respect to the new field direction C, and they carry out these jumps with probabilities $P(B_3, C_n)$. No atom can be found at a fixed angle of orientation with respect to both directions, B and C, that is in a state B_3 and C_2 simultaneously in a *reproducible* fashion. That is, states of orientation with respect to field directions B and C are mutually *incompatible*. Similarly, the state of possessing a certain x-co-ordinate and the state of possessing a certain x-momentum p_x are incompatible. Likewise, energy states and position states are incompatible. That is, an energy meter may have ascertained a certain energy value E_3 of a particle; a subsequent position measurement yields the x-co-ordinate x_5; if the energy test is now repeated it will not yield the former E_3 again with certainty, but may yield as well another energy value. The fact that mechanical observables, such as E and x are mutually incompatible is of fundamental importance for quantum mechanics.

'Mutually orthogonal states' of a system belong to mutually exclusive values of observable quantities or 'observables'. For example, A_1 A_2 . . . may signify those states of a particle in which its energy is E_1 E_2 . . . The states B_1 B_2 . . . may be states of position at a certain time with co-ordinates x_1 x_2 . . . . Whether the characteristic values or 'eigenvalues' of E and of x disclosed by E- and x-meters represent a discrete or a continuous set of values, is irrelevant in this connection, although it is of great practical importance. (Electrons in atoms have a discrete set of negative energies and a continuous band of positive energies.) Furthermore, the same energy meter which ascertains values E, also ascertains values of the observable $E^* = E^3$; it thus could as well be denoted as an E^*- or an

E-cube meter. This is to show that a set of states *A* is associated not only with one observable *a* but also with other observables $a^* = f(a)$ where *f* is any unique function.

As to reproducibility of a measuring result of a state, notice that a first *x*-test may yield the result x_1, and another *x*-test later may yield another position x_2. This merely shows that *x*-values are not reproducible, that they do not define 'states' in the sense introduced by quantum theory, namely *reproducible or well-defined states.* Position *x* of a particle is not a (well-defined) state, but position *x* at a certain time instant *t* is. The set A_1 A_2 . . . may signify space-time location at time t_A, namely (x_1, t_A), (x_2, t_A), . . . , and the set B_1 B_2 . . . may signify states (x_1, t_B), (x_2, t_B), . . . The *B*-measurement (x at t_B) thus is not a 'repetition' of the *A*-measurement, and an *A*-result will not be reproduced with certainty in a *B*-measurement. *A*-results are connected with the *B*-results by definite probabilities $P(A_k, B_j)$, however. This example shows that, in case of non-conservative observables such as position, the instant of observation is essential for characterizing the 'state'. When saying that 'well-defined' or 'reproducible states' are those states which fit into the schema developed above, this is not a circular definition. On the contrary, a general survey of the experimental situation encouraged the founders of the theory, in particular Dirac([2]) first to establish the formal schema, applicable to unspecified objects ('atoms'), to unspecified situations ('states') connected by *P*-relations ('probabilities'). Afterwards it was found that this abstract schema corresponds to a physical model when its mathematical symbols and verbal expressions are *properly* associated with phenomena under *proper* operational definitions. And here it must be emphasized that position in space-time defines a 'state', whereas position alone does not. The physicist who carries out one experiment *after* the other is of course primarily interested in states at various times; and of particular interest to him are states conservative in time, such as energy states, angular momentum states, etc. The general schema, however, is independent of such specifications; it applies to conservative as well as to non-conservative well-defined or reproducible states. The fixation upon the viewpoint of a temporal sequence of states, erroneously described as the meaning of the Schrödinger time-dependent wave equation (section 32), obscures the broader significance of the quantum theory as a schema of probability relations between 'states', always meaning reproducible states.

Next let us discuss the concept of a certain 'kind' of system (atom). In the widest sense it is defined as a system composed of a certain number of elementary particles held together by mutual forces. There are also various 'restricted kinds' of an atom which we may study in the example of the hydrogen atom. This 'kind' of atom is capable of an infinite number of states with various internal and translational energies. Of more interest is a more 'restricted kind' of H-atom which is in one definite translational state, for example, with its centre of gravity at rest, and which still has a variety of internal states. One orthogonal set of internal states—we may denote it as the set A—is characterized by three quantum numbers nlm where n characterizes the energy, l the angular momentum, and m the component of the latter with respect to the fixed direction A of a magnetic field. (We disregard the spin.) Another set B of states of the same restricted kind of H-atom is characterized by quantum numbers nlm^*, where m^* refers to orientation with respect to a magnetic field of another direction B. Still another set of states is represented by various triples of 'parabolic' quantum numbers $n_1 n_2 m$ for states oriented in an electric field; and so forth. Each of these orthogonal sets of states has the same infinite multiplicity, $M = \infty$. Next consider a more restricted 'kind' of H-atom, for example, one with a definite energy E_4 with quantum number $n = 4$ once and for all. It still is capable of a variety of states, one set being characterized by pairs of quantum numbers lm, another by lm^*, and so forth; each of these sets has the same multiplicity, $M = 16$ (since l can have values 0, 1, 2, 3, and m varies from l to $-l$).

A still more restricted kind of H-atom is one with $n = 4$ and $l = 2$, say. Its orthogonal sets of states are characterized by quantum numbers m, or m^*; they all have the same multiplicity, $M = 5$. These examples bear out the theorem that the various orthogonal sets of states of a certain 'kind' of atom have *common multiplicity* M, yielding quadratic P-tables with M rows and M columns.

12. *Do states change without tests?*

Suppose a gas of N particles is found in a well-defined state A with particle positions $r_1 \, . \, . \, r_N$ at time t_A. At a later time t_C another position test is carried out yielding the positions $r''_1 \, . \, . \, r''_N$. The question as to what happened to the positions of the N particles between t_A and t_C is answered with 'nothing at all', for two reasons. *First*, any possible answer (the system was in this or that state $r'_1 \, . \, . \, . \, r'_N$ at an intermediate instant t_B) has no concrete meaning unless it is substantiated. A test at the intermediate instant may indeed find the particles in certain positions $r'_1 \, . \, . \, r'_N$; but this result would produce a new probability situation for the final configuration, differing from the one actually observed at t_C when no intermediate test has been carried out. *Second*, how could the system possibly have gone through any set of intermediate states on its way to, or 'aiming' at, the final state when the latter depends on the final testing instrument of which the system does not 'know' anything before it is actually tested. A gradual development of a system from an initial to a final state through a continuity of intermediate states is alien to the quantum theory, since the final state is conditioned by the instrument which ascertains it. This viewpoint cannot be emphasized too much.

The statement 'nothing happens to the *state* of a system between two tests' has been challenged by the following example of Schrödinger.(3) A box contains a flask with poison, and a radium atom which, if and when it disintegrates, will open the flask by a trigger mechanism. The box also contains a cat. At time t_A the state inside the box is such that the flask is closed and the cat is alive. A test at time t_C shows the flask open and the cat dead. Are we really to believe that only the test at time t_C brought about the change of state, that the content of the box has not gone through a continuity of intermediate states, even without our looking or 'testing'? The answer is: 'You are using the term "state" in a different way as the quantum theorist who means "well-defined reproducible state" when he speaks of a state. In the present case, a well-defined position test at t_A would have to explore the exact positions of all elementary particles within the box at time t_A, which alone would have killed the cat. What you call state is only a rough estimate of the distribution of matter, organic and inorganic, within the box. Ill-defined macro-states indeed develop continuously without looking at them. Your example is tricky in so far as you also include a micro-physical radium atom which is in a well-defined state, either intact, or disintegrated. This introduces an element of uncertainty with betting odds for "cat alive"; these odds decrease from unity to zero with increasing time.' Macro-states are 'between' other macro-states. But there are no intermediate well-defined micro-states interposed between an initial and a final well-defined micro-state. It must be conceded, however, that the connection between micro- and macro-physics is in urgent need of clarification.

13. *Measurement*

Niels Bohr(4) in his essay, 'On the Notions of Causality and Complementarity', writes: 'In this situation we are faced with the necessity of a radical revision of the foundation for the description and explanation of physical phenomena. Here it must above all be recognized that, however far quantum effects transcend the scope of classical physical analysis, the account of the experimental arrangement must always be expressed in *common language* supplemented by the terminology of classical physics. This is a simple logical demand, since the word "experiment" can in essence only be used in referring to a situation where we can tell others what we have done and what we have learned.'

I have followed Bohr's 'logical demand' by using as testing instruments in the previous discussion only such arrangements, called 'filters' or 'separators' or 'meters' which are not appreciably influenced by the tested objects. They are *macrophysical instruments* which remain intact even when they throw the tested micro-object from its original state to one of the states characteristic of the 'meter'. For example, a double-refracting crystal is a 'separator' for two orthogonal states of polarization A_1 and A_2 whereas a Nicol prism is a 'filter' which passes A_1 and repels A_2. The Nicol itself, however, is supposed not to be affected by its own activity. Another example: What is commonly known as a yardstick, is a 'position separator' or 'position meter' which can be used over and over again to yield objective reproducible positions of a particle, relative to a fixed 0-point. Similarly, an energy meter is supposed to be so massive as to record energies of a particle without being affected itself; and so forth. This simple theory of measurement may, at a more advanced stage, be generalized into a theory of interaction between a small object and a small instrument. However, the term 'measurement' (or 'preparation of a state', Margenau) ought to be reserved for a 'situation where we can tell others what we have done and what we have learned' in an objective confirmable sense. However, if 'states' are reproducible, then I cannot see why one should need a new *philosophy* of knowledge in which the word 'objective state' is eliminated. Quantum physics has shown merely that we ought to be careful with applying the term 'state'. Position q is a state, and so is momentum p, but there is no combination (q, p)-state. Similarly there are states of viscosity v of water, and states of twist w of the same sample of water in the frozen state. But, although there are no reproducible combination (v, w)-states, nobody has found it necessary to introduce a new philosophy of knowledge. In the writer's opinion, the much-admired quantum philosophy concerning the failure of 'objectivity' of states is misleading by exaggerating trifles; and it does not even heed Bohr's 'logical demand'. Q-states, p-states, v-states, and w-states are reproducible, hence 'objective', whereas pq-combinations and vw-combinations are not reproducible, hence do not even deserve the epithet 'state'. It is true that the incompatibility of p and q, in contrast to the well-known incompatibility of v and w, has been a most important physical discovery. But if the vw-case does not call for a new philosophy of knowledge or a new language, then the pq-case does not either. More about this in Chapter V.

14. *Solution of the Gibbs paradox*

At first sight it may seem paradoxical that the postulate of continuity should be at the root of a branch of physics notorious for its discontinuous events. However, it is not continuity as such but *cause-effect continuity* which entails discontinuous events (section 7); and it is significant therefore that the discontinuous 'jumps' from state to state do *not* obey deterministic cause-effect relations. The apparent antinomy thus dissolves. It is interesting in this context to observe that the original quantum hypothesis of discontinuity, 'an oscillator of frequency ν can change its energy only in discrete amounts $h\nu$', was forced upon Planck when he tried to restore continuity to statistical thermodynamics, as follows.

According to classical statistical mechanics, the *kinetic* energy of temperature motion within a solid body (or a radiation aggregate) consisting of oscillators is distributed in equal shares, $\frac{3}{2}kT$, over the various constituting particles (or oscillators), irrespective of whether such a particle (oscillator) is tightly bound with high vibration frequency ν or loosely bound with low frequency ν; only entirely fixed particles (frequency $\nu = \infty$) are excluded from the *equipartition* of the thermal energy. This result of classical mechanics involves an obvious discontinuity, however. Suppose the force-constant binding a particle is increased; its thermal energy of motion remains $\frac{3}{2}kT$. Only in the last moment when the particle becomes entirely fixed, will its thermal energy abruptly decrease to *zero*. Observation, however, in agreement with the continuity postulate shows a *gradual decrease* of the thermal energy per particle with increased magnitude of its force-constant, that is with increased frequency of its vibrational motion. Planck overcame the discrepancy between classical theory and experiment, at the same time re-establishing *energy continuity* by introducing the unclassical hypothesis that the energy of an oscillator can only be a multiple of the quantum $h\nu$. At low temperature, then, a tightly bound particle, or an electromagnetic vibration of high frequency ν, can climb to the high-energy level $h\nu$ only at rare occasions, and will dwell mostly in the level $E = 0$; thus it will exhibit a low average energy *gradually* approaching zero when ν gradually approaches infinity. Planck's hypothesis was a stroke of genius; it was the one correct assumption, among many possible ones which would have restored continuity of the thermal energy. Yet it was an *ad hoc* hypothesis which could not have been deduced from the postulate of energy continuity alone.

The postulate of entropy continuity(5) which we are going to introduce in the following does not suffer from this defect. Without needing further intuition it leads straight to those phenomena which are illustrated in Fig. 1*c*. Entropy has always been defined with the help of 'filters' or semi-permeable diaphragms. Yet classical statistical mechanics suffers from a discontinuity pointed out first by W. Gibbs in the 1870's. In order to illustrate the *Gibbs entropy paradox*, imagine two equal quantities of the same kind of gas particles at the same temperature in two equal volumes V separated by a wall. The particles of the one gas may be in a state A, those of the other gas in the state B. For example, the particles may be silver atoms, and the atomic axis directions of gas A may point towards the north, those of gas B at an angle α away from the north. (The axis directions may be conserved due to extreme dilution.) If we now remove the wall between the two volumes, either gas will spread over the combined volume $2V$. This diffusion process is connected with an increase of the entropy S and of the maximum isothermal work W which can be gained by letting the diffusion process take place in an isothermal reversible way, for example with the help of a semi-permeable diaphragm which passes the one gas freely, but suffers a recoil pressure from the other gas. The maximum isothermal work δW and the entropy increase δS in the process of diffusion are

$$\left.\begin{aligned} \delta W &= 2NkT \,.\, ln2 \\ \delta S &= 2Nk \quad .\, ln2 \end{aligned}\right\} \text{ for } A \neq B \tag{5}$$

where $2N$ is the total number of gas molecules in the two samples. On the other hand, if the two states A and B are *equal*, work cannot be gained from the diffusion process, so that in this case

$$\left.\begin{aligned} \delta W &= 0 \\ \delta S &= 0 \end{aligned}\right\} \text{ for } A = B \tag{6}$$

Imagine now that the difference between A and B, for example the angular difference of the atomic axes, is gradually decreased and finally is made to disappear altogether. As long as there is a difference, however small, the mixing of the two gases, according to classical ideas, yields the full maximum isothermal work and entropy increase (5). If finally the difference (for example the angle α) is decreased from an extremely small finite value to exactly *zero*, then, according to classical statistical thermodynamics, δW and δS decrease abruptly from the value (5) to *zero* as in (6). This is the *discontinuity paradox* of Gibbs, which classical theory cannot solve, although various answers have been proposed. For example, it is said that the paradoxical situation envisaged by Gibbs never occurs in nature since two gases either have a finite difference, or no difference at all, since gas differences are determined by integral numbers, for example by the number of electrons in the molecules, etc. Therefore, *do not worry about the paradox*!

I call this 'disputing away' the paradox rather than solving it. It is true that different *kinds* of gas atoms or molecules cannot be made equal in a gradual way. This argument fails, however, for different *states*, A and B, of the same kind of particles. The considerations of Chapter I have shown that the general cause-effect continuity postulate requires the splitting effect of Fig. 1*c*, hence the existence of fractional equalities between two states, $A \sim B$, as intermediate between $A = B$ (inseparability) and $A \neq B$ (separability). In the case $A = B$ the B-particles penetrate the A-filter, hence they do not exert pressure on it. In the case $A \neq B$ the B-particles produce pressure. In case of $A \sim B$ only the repelled B-particles exert pressure and contribute to the δS-value. When the state of the particles is gradually changed from $B = A$ to $B \neq A$ via $B \sim A$, the diffusion entropy δS gradually changes from the value 0 to $2Nkln2$.

Take the example of silver atoms in various orientations of their magnetic axes. The idea of a semi-permeable diaphragm or filter which passes N-directed atoms but blocks those with axes ever so little deviating from the N-direction is 'unphysical'. In fact, a Stern-Gerlach field of N-direction may serve as a 'separator' passing both N-directed and S-directed atoms. When blocking the path of the S-atoms, the separator becomes a 'filter' passing N-atoms only. This N-filter reacts to incident atoms deviating by an angle α from N by passing the fraction $\cos^2(\frac{1}{2}\alpha)$ and blocking the fraction $\sin^2(\frac{1}{2}\alpha)$. Only the latter fraction exerts a gas pressure on the filter. A gradual change of α thus leads from the full entropy of diffusion, $\delta S = 2Nk \,.\, ln2$, to $\delta S = 0$ when α is changed from 180 deg. to zero. *The Gibbs paradox is thus solved*: it is removed by virtue of the splitting effect of Fig. 1*c*, usually regarded as a quantum effect.

In a previous book of 1955 quantum theory was developed on a thermodynamic basis, by first establishing the principle of *entropy continuity* which maintains that 'the Gibbs paradox does not occur'. This leads to postulating intermediate δS-

values between 0 and $2Nk \,.\, ln2$, that is, fractional pressures of B-particles on A-filters, hence the splitting effect of Fig. 1*c* which is already one of the principal quantum features. However, since Boltzmann, mechanics comes first, and thermodynamics has second rank as a statistical consequence. Therefore, it is preferable to establish quantum theory on a mechanical basis, as done in this book. The solution of the Gibbs paradox of thermodynamics, then, is but a special implication.

15. *Entropy increase by testing*

Suppose N independent gas particles of the same kind are all in the same original state 0. When the gas is now subjected to an A-separator test, it will emerge as a mixture of $n_1\, n_2 \ldots n_M$ particles in the states $A_1\, A_2 \ldots A_M$ respectively. The mixture has added entropy:

$$S_A = N\, ln\, N - \Sigma_k\, n_k\, ln\, n_k \tag{7}$$

which is larger than the original entropy

$$S_0 = N\, ln\, N - N\, ln\, N = 0$$

The resulting gas mixture may now be subjected to a B-separator test; its particles will then jump to the new states $B_1\, B_2 \ldots B_M$, and the occupation numbers of the states B will be

$$m_j = \Sigma_k\, n_k\, P\,(A_k, B_j)$$

yielding a new entropy

$$S_B = N\, ln\, N - \Sigma\, m_j\, ln\, m_j \tag{8}$$

Von Neumann[6] has given a mathematical proof on grounds of the unit sum rules (II 3*a*, *b*) for the P-tables, that S_B is larger than S_A. Similarly, S_C is larger than S_B, and so forth; thus test after test produces an ever increasing entropy

$$S_0 < S_A < S_B < S_C \ldots \tag{9}$$

The maximum of S is obtained when finally a state of equipartition over the M states of an orthogonal set is reached. In classical theory a gas of low entropy is supposed to show larger and larger entropies as seen in subsequent ***inspections*** from outside. Quantum theory yields the same result under subsequent *tests* by means of separators which infringe on the gas by redistributing the particles over new sets of states.

One must not forget, however, that statistical results such as (9) hold only for the overwhelming majority of cases; actually there are individual fluctuations away from the aver age. Thus although the entropy of a gas mixture will most probably increase by virtue of the next separation test, it may actually decrease, and under a long series of tests, the entropy will fluctuate up and down near the entropy maximum, as illustrated by the up and down staircase curve for S as a function of time t.

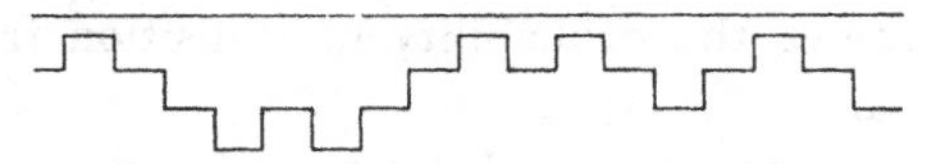

Fig. 4. Entropy S as a function of time t.

This curve (Fig. 4) was first drawn by P. and T. Ehrenfest in order to illustrate the entropy of a gas according to *classical statistical mechanics*; the various staircase levels were to signify S-values obtained in a series of inspections of the gas from outside. Yet, whether one assumes classical inspection, or quantum infringement by testing, the fluctuating S-curve is *symmetric* with respect to the positive and negative time direction. In other words, if the S-curve is shown in a film, one cannot tell whether the film is run forwards or backwards.(7) The doctrine of a parallelism of time and entropy is untenable, in classical as well as in quantum theory.

16. *Summary*

(*a*) The purely formal content of Chapter II may be condensed as follows. The elements of a class S of entities are supposed to be connected by positive quantities $P(S_l,S_j)$ less than unity, with only $P(S_l,S_l) = 1$. The elements of S can thereby be divided into subclasses A, B, C, . . . so that each subclass contains a complete set of mutually orthogonal elements with $P(A_k, A_{k'}) = \delta_{kk'}$. The P-fractions connecting elements of different subclasses are supposed to satisfy the unit sum rule, $\Sigma_j P(A_k, B_j) = 1$, for the rows of the P-matrices. The P's are to be symmetric, $P(A_k, B_j) = P(B_j, A_k)$. Hence, the unit sum rule holds also for the columns, $\Sigma_k P(A_k, B_j) = 1$, so that the P-matrices are 'unit magic squares'. That is, the various subclasses A, B, C, . . . have common multiplicity.

(*b*) The formal substructure becomes a physical theory by identifying the entities S with the (well-defined) states of a mechanical system, and $P(A_k, B_j)$ with the fractional equality degree between the states A_k and B_j defined as a passing fraction in a test. Under the postulate of reproducibility, $P(A_k, B_j)$ becomes the probability of transition from A_k to B_j or vice versa.

(*c*) The motivation (or explanation) of the mathematical schema and of its physical interpretation may be seen in the empirical postulates of *cause-effect continuity*, bridging the gap between equality and inequality, that is separability and inseparability. The postulate of *reproducibility* of a test result then necessitates transitions from state to state in filter tests. The postulate of a two-way *symmetry* for transition probabilities corresponds to the classical reversibility of deterministic processes.

CHAPTER III

THE METRIC LAW OF PROBABILITIES

Wave-Particle Duality?
'Je n'ai pas besoin de cette hypothèse.'

LAPLACE

Someone experimenting with paper triangles finds that the three altitudes in every triangle intersect in one point, and that the same holds for the three angular bisectors, the three medians, and the three perpendiculars on the side-centres. He concludes that he has discovered a new general principle of 'unity in triplicity' which is confirmed not only in geometry but also in music, architecture, political institutions, and so forth. A mathematician will tell him, however, that his geometrical unity in triplicity is not a separate 'principle' in its own right, but rather is a consequence of the Euclidian axioms, and can thus be understood as a geometrical necessity, instead of being accepted as a cabbalistic oddity with deep philosophical implications. Similarly, the empirical fact that the statistical behaviour of particles is wave-like may certainly be described by the comprehensive term 'wave-particle duality'. However, when physicists after half a century of quantum physics still believe that duality, together with a complementarity of particle- and wave-features, reveals an independent fundamental 'principle' of nature, an immanent trait of the microcosm with ramifications in other domains of human knowledge([1]), then it is time to pause and reflect whether this duality might not be reducible to simple, almost self-evident physical ground axioms, so that we can recognize it as a necessity rather than an oddity. I think indeed that simple and elementary physical postulates can be established which explain why the statistical manifestations of particles should obey a law of 'interference of probabilities' *via* a complex-imaginary amplitude called ψ, and why $E = h\nu$ and other quantum rules prevail, that is why the co-ordinates and momenta of particles should be in a periodic wave-like relation. Such a reduction from the 'strange' to the 'natural' is all the more desirable as the alleged mysterious character of the quantum rules seems to be generally taken for granted, as underlined, for example, when P. Bridgman(2) contends: 'When we go far enough in the direction of the very small, quantum theory says that our forms of thought fail, so that it is questionable whether we can properly think at all,' and when even Heisenberg(3) declares that the quantity ψ is 'abstract and incomprehensible . . . so to speak containing no physics at all'. In order to dispel the aura of incomprehensibility surrounding the esoteric quantity ψ let us first consider a geometrical analogy.

17. *The metric of geometrical structures*

Suppose an infinite number of sticks, each labelled with two terminal letters, are to be cut to such lengths L_{AB}, L_{AC}, L_{BC}, . . . that the L's are correlated by a general mathematical law which is symmetric in the terminal letters, self-consistent, self-reproducive, and in short has group character, under the conditions

$$L_{AB} = L_{BA}, \text{ and } L_{AA} = 0.$$

In order to illustrate what I mean by a correlation law of group character, suppose three quantities—let us call them ψ_{AB}, ψ_{BC} and ψ_{AC}—are connected by a functional relation

$$\psi_{AC} = f(\psi_{AB}, \psi_{BC}),$$

so that its inversion expressing ψ_{AB} in terms of ψ_{AC} and ψ_{CB} yields

$$\psi_{AB} = f(\psi_{AC}, \psi_{CB})$$

with the same function f again. Furthermore, one also wants to obtain, always with the same function f

$$\psi_{AC} = f(\psi_{AD}, \psi_{DC}) = f(\psi_{AE}, \psi_{EC}) = \text{etc.}$$

These requirements, in particular that f_{AC} is to be 'independent of the path', can be satisfied (apart from 'distortions') only† when the function f indicates addition:

$$\psi_{AC} = \psi_{AB} + \psi_{BC} \text{ together with } \psi_{AB} = -\psi_{BA} \qquad (1)$$

This triangular correlation law has group character indeed, that is one obtains ψ_{AC} not only via B but also via other intermediate letters when replacing ψ_{AB} by $\psi_{AD} + \psi_{DB}$, and ψ_{BC} by $\psi_{BE} + \psi_{EC}$, etc. The symmetric quantities L, for which a general correlation law is sought, can of course not be identified with the anti-symmetric quantities ψ of (1). However, with (1) as a substructure one now may construct L-correlation laws of group character in the following way.

Interpret the quantities ψ as *vectors* connecting points $A, B, C, ..$ distributed either in one- or two- or higher dimensional space. Then interpret the L's as the distances:

$$L_{AB} = |\psi_{AB}| = |\psi_{BA}| = L_{BA}$$

The resulting L-correlation laws depend on the dimension of the point structures as follows:

(a) If the points $A, B, C, ..$ are distributed in one dimension on a straight line (or on a circle) then two lengths L_{AB} and L_{BC} determine L_{AC} only in a bivalent fashion since C may be located on either side of B and A. However, the six connections L between four points A, B, C, D are in a fixed relation so that five L's uniquely determine the sixth L, by a general mathematical precedure which is symmetric in the letters and has group character, that is, can be continued to any other set of points. The L's are in a *tetragonal relation*, one-dimensional structures being supported by four points.

(b) If the points are on a plane (or a spherical surface) then one has a *pentagonal* relation between the L's. Five points are connected by ten distances L. When nine of them are given, they determine the tenth L by a unique correlation law which is symmetric in the letters A, B, C, D, E and holds for any five letters.

(c) In three dimensions there is a hexagonal L-relation law. Six points are connected by fifteen L's. Fourteen of them uniquely determine the fifteenth.

And so forth for higher dimensional spaces, flat or of constant curvature. Although the L-correlation laws become more and more involved with higher dimensions, the simple vector law (1) holds in any number of dimensions. In the special case (*b*) of structures in a *plane* one can denote the vectors ψ by complex symbols, $\psi = L\, e^{i\phi}$, without thereby imbuing them with

† The distortion $\omega = e^{\psi}$ leads for ω to $\omega_{AC} = \omega_{AB}\, \omega_{BC}$, together with $\omega_{AB} = 1/\omega_{BA}$.

any abstract or incomprehensible quality. In order to obtain a closer analogy to the probability theory developed in the next paragraph, one may denote L^2 by the letter P and thus write $P = |\psi|^2$.

It is an interesting mathematical question whether the P- and L-correlation laws obtained above via the ψ-addition law as a substructure are the only possible correlation laws of group character. Can one construct other general L-correlation laws not resting on (1)? Or can one give a *uniqueness proof* for the procedure above as the only possible one? (Refer to Appendix 2.)

18. *The metric of probability amplitudes*

We now turn to the question of constructing a set of unit magic squares, (P_{AB}), (P_{AC}), (P_{BC}), . . . in such a manner that they are connected by a general self-consistent correlation law of group character, symmetric in the letters A, B, . . . (which indicate sets of states A, B, . . . of a physical object). Maybe one could construct a trigonal correlation law so that (P_{AC}) is uniquely determined by (P_{AB}) and (P_{BC}). Or one may think of tetragonal, or pentagonal, etc., laws, in analogy to the geometrical cases (*a*), (*b*), (*c*), etc., of section 17. At any rate, the P-matrix correlation law is required to maintain the unit magic square quality unchanged or 'invariant' under the 'transformation' which expresses one P-table in terms of other P-tables.

A mathematician asked to establish such a correlation law between unit magic squares (refer also to Appendix 2) will remember that the required invariance is satisfied by the well-known law of *unitary transformation* which runs as follows. Associate with every positive quantity $P(A_k, B_j) = P(B_j, A_k)$ two complex conjugate quantities

$$\psi(A_k, B_j) = \sqrt{P}e^{i\phi kj} \text{ and } \psi(B_j, A_k) = \sqrt{P}e^{i\phi jk}$$

with opposite phases, $\phi_{jk} = -\phi_{kj}$, left undetermined so far. Every P-table, e.g. (P_{AB}), is thus associated with a ψ-table:

$$(\psi_{AB}) = \begin{bmatrix} \psi(A_1, B_1) & \psi(A_1, B_2) \ldots \\ \psi(A_2, B_1) & \psi(A_2, B_2) \ldots \\ \ldots & \ldots \quad \ldots \end{bmatrix}$$

Now choose the phase angles ϕ of the quantities ψ (that is the directions of the vectors ψ in a common plane) in such a manner that the ψ-matrices will satisfy the triangular matrix multiplication law:

$$(\psi_{AC}) = (\psi_{AB}) \cdot (\psi_{BC}) \tag{2a}$$

made self-consistent by

$$(\psi_{AA}) = (\psi_{AB}) \cdot (\psi_{BA}) = (1) \tag{2b}$$

Written out in detail this is

$$\psi(A_k, C_n) = \Sigma_j\, \psi(A_k, B_j) \cdot \psi(B_j, C_n) \tag{3a}$$

made self-consistent by requiring as a special case of (3c):

$$\psi(A_k, A_{k'}) = \Sigma_j\, \psi(A_k, B_j) \cdot \psi(B_j, A_{k'}) = \delta_{kk}{'} \tag{3b}$$

From this ψ-substructure one now arrives at unit magic square P-tables by defining $P = |\psi|^2$, that is

$$P(A_k, B_j) = \psi(A_k, B_j) \cdot \psi(B_j, A_k) \tag{4a}$$

In order that the left-hand side be real positive, the two factors on the right are to be complex conjugates, as remarked before:

$$\psi(A_k, B_j) = \psi^*(B_j, A_k) \tag{4b}$$

Equation 3*a*, known to mathematicians as the law of *unitary transformation*, is identical with the law of *interference of probabilities* of quantum theory. As a complex quantity each ψ gives a direction in a plane to the corresponding P.

The interference of probabilities *via* the ψ-law (2*a*) has always been regarded as one of the most puzzling features of micro-mechanics, as 'beyond the understanding of the physicist . . . who must be content to accept the implications without hoping to penetrate the mystery that is implied' (Lennard-Jones).([5]). The esoteric nature of ψ is further underlined by Niels Bohr([6]) who holds that 'these symbols themselves, as is indicated by the use of imaginary numbers, are not susceptible to pictorial interpretation'. However, a vector structure in a plane, with each probability having a direction, is quite pictorial. And 'the mystery that is implied' is merely the fact that the various probabilities are not chaotic but are connected by a general correlation law which, under the circumstances of having to connect unit magic squares, must be that of unitary transformation. This law has a formal analogy to wave theory, in particular when the directional angles of the ψ-vectors are re-interpreted as 'phase angles'. It seems rather far-fetched, however, to take this formal analogy for a justification of the metaphysical belief that nature is dualistic, that interference of probabilities is a wave-like law imposed on particles by some whim of nature. There is no wave-particle duality; there rather is a *unitary particle theory* in which the appearance of 'wave interference' (and of qp-periodicity, Chapter IV) can be *explained* on the basis of simple non-quantal postulates. There is hardly a simpler and more 'natural' (though in the last resort metaphysical) postulate than that there ought to be a general structural law connecting the various probabilities P arranged in unit magic squares.

The question arises, however: Is unitary transformation (2 *a*) the *only possible* way of establishing a general correlation law between unit magic squares? Can one construct other correlation laws? If not, can one devise a *uniqueness proof* that, if there is to be a general P-relation law at all, then unitary transformation is the only answer? Suppose a uniqueness proof can be given. Only then can 'interference of probability amplitudes' be regarded as a necessity without qualification. Still, it seems *inconceivable* that another general P-correlation law could be constructed with results differing from those of unitary transformation. Therefore, one may even now regard interference of probabilities as a 'natural' law which can be understood, without resorting to a separate 'principle' of wave-particle duality. The situation here is similar to that in statistical mechanics where one accepts the derivation of thermodynamical facts from a statistical basis even in the absence of a rigorous proof of ergodicity. Still, a conscientious reader may agree with Eddington's remark made on another occasion: 'The building at this point shows some cracks; but I think it should not be beyond the resources of the mathematical logician to cement them up.'

19. *The metric of probabilities*

The formal analogy between the ψ-vector addition law (1) and the ψ-multiplication law (2*a*) goes even further. As mentioned before, when five points A, B, C, D, E in a *plane* are connected by ten distances L, or ten square distances P, nine P's uniquely determine the tenth P. Similarly, it can be shown that the triangular ψ-matrix law (2 *a*) is equivalent to a direct relation law between ten P-tables which connect any five orthogonal sets of states A, B, C, D, E so that nine unit magic squares uniquely determine the tenth.

This mathematical result has interesting implications. From the very beginning plane geometry has consisted mainly in exploiting special cases of the unique relationship between ten lengths in plane structures connecting five points. Only around the year 1900 did the vector law (1) come into general usage without, however, contributing any new insight into geometrical relations. The development of probability theory in physics has taken the opposite path. Here one first has discovered, after a long path of trial and error, the triangular product law (2*a*) for matrices of complex quantities ψ, that is probability vectors in a plane. The same ψ-law leads to, and is equivalent to, a direct law between ten P-matrices (see above). The moral is that the interference law of probability amplitudes ψ is no more and no less essential for the interconnection between transition probabilities P than the belated introduction of vectors has been for geometry in a plane. Of course, nobody will overlook the great practical value of the ψ-law for finding quantitative relations between probabilities P in those cases where the phase relations of the corresponding amplitudes ψ happen to be known from independent souces. [They are known in particular for the standard function $\psi(q, p)$, Chapter IV.] One must not forget, however, that the phases, that is the directional angles, occurring in a single matrix (ψ_{AB}) are irrelevant for the probabilities in the single matrix (P_{AB})—just as the direction of a single rod is irrelevant for its length. Only when one wishes to describe relations between at least three P-tables (P_{AB}), (P_{BC}), (P_{AC}), or between three rods in a triangle, then and only then do the ψ-vector laws become operative.

If one should restrict the quantities ψ to the *real* values $\psi = \pm\sqrt{P}$, that is if one allows only two opposite directions along a line, 'unitary transformation' would then become 'orthogonal transformation'. The resulting P-tables would still be unit magic squares; but they would belong to a restricted class of such squares, too narrow for the requirements of quantum mechanics. The reason for admitting complex probability amplitudes, that is vector directions in a *plane*, rather than only along a line, will become clear in Chapter IV.

Summing up: whereas the relation between probabilities of transition via interference of amplitudes ψ is usually introduced as an *ad hoc* rule, we have seen that the same ψ-law is of an almost aprioristic geometrical character; it is the one and only conceivable way of satisfying the requirement that the P-tables be connected by a self-consistent general interrelation law at all. With this result it is not necessary any more to appeal to a 'principle' of duality and complementarity in order to justify the wave-like behaviour of particles.

On this occasion I wish to contradict the view that interference of probabilities is extraordinary because according to ordinary reasoning one would have to expect the probabilities P to obey a law which actually, according to (3*a*), holds for probability amplitudes ψ. I have never understood why such a P-law should hold according to 'ordinary' expectation. Suppose A_k and C_n are two states of position ten feet apart, and halfway between them are the positional states

$B_1 B_2 \dots$ A cat can easily jump five feet, but only rarely jumps as far as ten feet. Why, then, should an 'ordinary' cat arrive directly from A_k at C_n with a probability which is the sum of all the probabilities of arriving at C *via* the various intermediate places $B_1, B_2, \dots$? Furthermore, a P-law of the form (3*a*) could never constitute a *general* law; it would fail in the simple case of C_n being identical with $A_{k'}$ for $k' \neq k$ where it would read

$$P(A_k, A_{k'}) = \Sigma_j \, P(A_k, B_j) \cdot P(B_j, A_{k'}) \quad \text{(wrong!)}$$

with *zero* on the left and a sum of positive terms on the right.

As seen before, an individual ψ-vector direction does not have physical significance of its own. The complex quantities ψ with their directions in a plane merely serve to establish a mathematical relation between various transition probabilities plotted in tables (P_{AB}), etc., after being observed in tests with macroscopic instruments, shortly called A-meters, B-meters, and so forth. This ought to dispose of the idea that a ψ-function or ψ-wave, which represents one row in a ψ-matrix, has any direct physical substantiality as a 'pilot wave' or 'quantum potential' which 'guides' events (de Broglie, Bohm, secs tion 30), or that it gradually expands and occasionally shrink-suddenly (Heisenberg, section 29). A ψ-wave does not guide actual events any more than a mortality table guides actual mortalities, and it shrinks no more than a mortality table shrinks when an actual death occurs. Material waves were in order thirty years ago; in fact they inspired the founders of wave mechanics to establish their marvellous method of calculating atomic data. Today the same ideas obscure rather than clarify the significance of all that quantum theory stands for, which is: to provide us with formulas, tables, or other rules of correlation between events, in the present case between probabilities of transition and values of observables revealed in tests. But these rules ought to, and can be understood as *necessities* rather than as oddities under certain simple ground postulates. Referring to an article by C. Lanczos[4]:

'"God always geometrizes" said Plato. "God always geometrizes" said Kepler. "God always geometrizes" said Einstein. The dream of the old philosophers became reality, the universe was explained as a magnificent mechanism of lawful interactions, and the law itself became a manifestation of a supreme mathematical wisdom which operates in the universe.'

It has been known for a long time that the law of unitary transformation is at the root of 'interference'. But the general implication for natural philosophy is entirely missed if one clings to a 'fundamental wave-particle duality' as a principle in its own right.

There have been several ingenuous attempts to modify or generalize quantum theory by the introduction of new forms of matrix algebra in which the law of association, and that of commutation on a higher level, is dropped (refer to P. Jordan[7]). My question is always: do these mathematical developments represent new possibilities of correlating unit magic square tables? If not, then I cannot see too much chance that they will lead to progress in quantum physics, however interesting they may be from the purely mathematical point of view.

On the other hand, the fact that atomic events are dominated by the elementary correlation law of unitary transformation, whereas ordinary games with dice and roulettes are not, may be taken as a sign that the microphysical quantum games deal with truly fundamental events and really elementary particles rather than with complex mechanisms.

20. *Identical particles and Nernst theorem*

A topic belonging to the general probability metric independent of the dynamical constant h is the strange predilection of identical particles for crowding together in an 'unclassical' fashion in the same state (Bose-Einstein particles) or pushing one another out from the same state (Pauli-Fermi particles). The two modes of interaction are closely connected with the fact that the probability amplitued ψ as a function of the position, or of any other observable quantity pertaining to identical particles, is either *symmetric or anti-symmetric* (see below) with respect to permutations of the particles. The symmetry rules which are responsible for the structure of molecules and bodies in general, are said to involve 'some aspects of great philosophical content beyond the understanding of the physicist . . . who must be content to accept the implications without hope to penetrate the mystery that is implied' (Lennard-Jones)([7]). And the same 'mystery' has been one of the reasons why Einstein regarded quantum mechanics as an incomplete theory. Let us therefore make sure that the symmetry or anti-symmetry of ψ rests on a far from mysterious foundation, namely on the following self-evident *postulate*:

When identical particles form a system, then any observable quantity pertaining to the system will display the same value whether particle a is at the place x_1 and b at x_2, or whether b is at x_1 and a is at x_2.

From this follows that, whenever the two equal particles perturb one another, and when a has co-ordinates $x_1y_1z_1s_1$ and b has co-ordinates $x_2y_2z_2s_2$ (s means spin-coordinate with two possible values $+\frac{1}{2}$ or $-\frac{1}{2}$), then

$$\begin{array}{lll} \text{either} & \psi(a_1b_2) = \psi(b_1a_2) & (\text{symmetric } \psi) \\ \text{or} & \psi(a_1b_2) = -\psi(b_1a_2) & (\text{anti-symmetric } \psi) \end{array} \qquad (5)$$

The proof of this result is found in every textbook; it involves the elements of perturbation theory, that is it supposes that the two equal particles are in mutual 'interaction at a distance', so that they 'know' of each other's position and can thus regulate their common statistical behaviour in mutual accordance. There is nothing 'beyond the understanding of the physicist' in this result since it is a direct consequence of the self-evident postulate above in conjunction with the general ψ-formalism of unitary transformation or 'interference'. Yet the consequences of symmetry and/or anti-symmetry on the constitution of matter are tremendous.

Even more simple is the proof that, in case of three (or more) identical particles a, b, c, . . . in mutual interaction, ψ cannot be symmetric with respect to an exchange of a and b and at the same time anti-symmetric with respect to an exchange of a and c. Indeed, if it were so then one would have the sequence

$$\psi(a_1b_2c_3) = \psi(b_1a_2c_3) = -\psi(b_1c_2a_3) = -\psi(a_1c_2b_3)$$
$$= +\psi(c_1a_2b_3) = +\psi(c_1b_2a_3) = -\psi(a_1b_2c_3)$$

Since the result, $\psi(a_1b_2c_3) = -\psi(a_1b_2c_3)$ is self-contradictory, ψ can either be symmetric with respect to all three (or more) identical particles, or anti-symmetric with respect of all of them. This means, however, that the class of all particles is divided in two subclasses: those which form symmetric ψ's (Bose particles) and those which form anti-symmetric ψ's (Fermi particles). All this follows from the postulate that observable quantities ought to be symmetric with respect to any permutation of the identical particles, in short from the definition of identity as indiscernibility. It is far from self-evident, however, that the world contains uncounted billions of identical particles.

Symmetric and anti-symmetric interaction implies that particles at the end of a long molecule, or at any distance whatsoever, 'know' what their identical partners are doing at the other end, so that they will be able to do the same as much as possible (Bose-Einstein) or not to do the same at all (Fermi-Dirac). As to this mutual communication over long distances, one must not forget that non-relativistic mechanics takes it for granted that interaction, even over planetary and interstellar distances, is instantaneous. A relativistic theory of interaction between several charged particles is still in the exploratory stage, in ordinary as well as in quantum mechanics. A relativistic quantum theory would also have to explain why particles of half-integral spin always prefer the anti-symmetric mode of interaction. Pauli has recently made important forward steps towards a solution of this problem (refer to his article in *Niels Bohr and the Development of Physics*[2]).

The Nernst theorem, known also as the Third Law of thermodynamics, is connected with the symmetry or anti-symmetry of the ψ-functions of N identical particles forming a body. Every state of such a body could be realized by $N!$ ψ-functions leading to an $N!$-fold 'degeneracy', if it were not for the symmetry requirement which reduces the number $N!$ to only *one* permitted ψ-function, either symmetric or anti-symmetric.

The Nerst theorem must not be confused with the simple statement that one can never reach the absolute zero point $T = 0$. This empirical-looking statement becomes self-evident when one replaces the artificial absolute temperature scale T by the more natural scale $t = \log T$ where $t = -\infty$ stands for $T = 0$; and $t = -\infty$ can obviously never be reached. Chasing the absolute zero point is the Achilles chase in reverse. Achilles, of course, *reaches* the tortoise; only the trick of dividing a finite interval into an infinite number of steps makes it appear to be a surprising empirical fact that he actually succeeds. In contrast, the physicist of course *cannot reach* $t = -\infty$; only the trick of condensing the infinite t-interval into a finite T-interval makes it appear to be a surprising empirical fact that he actually will *not* succeed in reaching his goal. Considering the unattainability of $T = 0$ as a new law of thermodynamics is as misleading as regarding the fact that Achilles catches up with the tortoise as a new law of kinematics. Actually Nernst received the Nobel prize for a physical discovery about bodies approaching the (obviously unattainable) limit $T = 0$, namely, the experimental result that the entropies of various states of association and aggregation of particles forming a body converge towards a *common* entropy value differing by *finite* amounts from the entropies at higher temperatures. This is a very far-reaching empirical statement. Calling it the Third Law of Thermodynamics is not quite justified since it is a consequence of the fundamental symmetry principles as parts of the general ψ-metric.

21. *Summary*

(*a*) The only mathematically conceivable way of constructing a self-consistent interconnection law between unit magic square P-tables, pending a uniqueness proof, is the law of *unitary transformation*:

$$(\psi_{AC}) = (\psi_{AB}) \cdot (\psi_{BC}), \text{ where } P = |\psi|^2.$$

(*b*) With P interpreted as a probability of transition, unitary transformation is identical with interference of probability amplitudes ψ. The complex ψ's as vectors in a plane give direction to the corresponding P's. The triangular ψ-matrix law is equivalent to a direct pentagonal law for the P-matrices themselves, in perfect analogy to plane geometry where ten lengths L connecting the five points of a pentagon are always related so that nine L's uniquely determine the tenth L.

(*c*) Since unitary transformation is the only mathematically conceivable triangular law connecting probabilities arranged in unit magic square tables, the wave-like probability interference is not a manifestation of a separate quantum principle of wave-particle duality. This 'principle' may rather be explained as a mathematical necessity, if a general connection law between the transition probabilities (from A to B, from B to C, and from A to C) is supposed to exist at all, that is if there is to be order rather than chaos.

CHAPTER IV

QUANTUM DYNAMICS

'To tell us that every species of things is endowed with an occult specifick quality by which it acts and produces manifest effects, is to tell us nothing. But to derive two or three general principles of motion from phenomena, and afterwards to tell us how the properties and actions of all corporeal things follow from those manifest principles, would be a very great step in Philosophy, though the causes of those principles were not yet discovered.'

NEWTON

22. *Duality and conjugacy*

The most striking feature of quantum physics is the wave-like periodic relation between co-ordinates q and momenta p. and between energy E and time t, best known from the rules of Planck, $E = h\nu$, de Broglie, $p = h/\lambda$, and Bohr, $\nu = (E_1 - E_2)/h$. Modern quantum theory has replaced the earlier quantum rules by Born's qp-commutation and Schrödinger's p-operator rule, the latter two being equivalent to the periodic probability amplitude function

$$\psi(q, p) = \text{const}\exp(2i\pi qp/h) \tag{1}$$

that is a wave function in q-space of wave length $\lambda = h/p$, The earlier and now the modern quantum rules were introduced as *ad hoc* theorems in order to fit the observed facts. Shall we accept them simply because they 'work' so well, or could they perhaps be understood on the grounds of less sophisticated elementary postulates?

The traditional answer to this question has been most unsatisfactory. For the last thirty years the wave function (1) has been regarded as evidence of a dominant principle of nature according to which, using Newton's language, matter (and light) is 'endowed with an occult and specifick quality by which it acts and produces manifest effects', namely, a dualistic interplay of wave and corpuscular qualities, attenuated by as fundamental complementarity. And after so many years it is not even felt as problematic any more that matter sometimes displays particle, sometimes wave features. We are told that clicks in Geiger counters and Compton recoil effects can be explained only in corpuscular terms, whereas interference fringes in diffraction experiments can be accounted for only by wave theory. To quote from an otherwise excellent text-book:

'Electrons, instead of having laws similar to classical laws, obey the laws of wave motion, . . . and light is corpuscular in nature, at least when it interacts with matter.' It really looks like waves on Monday, Wednesday, and Friday, and like particles the rest of the week (Bragg). Only a few independent spirits, among them Einstein, Schrödinger, and de Broglie, have steadfastly refused to accept this 'quantum mess' as final.

The situation has hardly been clarified by the introduction of descriptive phrases such as 'duality'. With reference to Planck's and other quantum rules, von Weizsäcker([1]) writes: 'We know today that ($E = h\nu$) is a consequence of a basic fact of all atomic events, the dualism of the wave picture and the particle picture.'

Instead of accepting the 'occult and specifick quality' of dualism at face value, it is proposed in this book to 'derive it from two or three general principles of motion'. One of them has been the non-quantal postulate that the probabilities of transition, arranged in unit magic squares, are connected by some self-consistent general law (Chapter III); this law can then have but one conceivable form, namely that of unitary transformation, identical with a wave-like *interference* of probabilities. The other 'occult and specifick' quantum feature, the periodic relation between *conjugate* dynamical observables q and p, will be deduced in this chapter.

Conjugacy between quantities $q =$ co-ordinates and $p =$ momenta has been defined differently in classical and in quantum mechanics.

(a) *Classical mechanics* defines q and p as conjugate pertaining to a certain mechanical system when the equations of motion have the canonical form

$$dq/dt = \delta H/\delta p \text{ and } dp/dt = -\delta H/\delta q.$$

This definition refers to energy and time, which themselves represent a conjugate pair. The classical definition thus is circular.

(b) *Quantum mechanics* usually defines q and p as conjugate co-ordinates and momenta when they satisfy the Born commutation rule, $qp - pq = h/2i\pi$, or the Schrödinger operator rule, $p = (h/2i\pi)d/dq$, or when they give rise to the wave function (1). All three conditions introduce the quantum relation between q and p by decree, thus putting quantum dynamics on an *ad hoc* basis.

Our task is that of deriving quantum dynamics as a consequence of simple general postulates, and in particular obtaining the periodic relation between q and p without introducing special quantum rules. In contrast to the definition (b) above we require q and p only to satisfy the following demands:

(c) A physical quantity defined in terms of two (or more) values p_1 and p_2 of a linear (not angular) momentum component shall depend only on the *difference*, $p_1 - p_2$. That is, there is no preferred zero point in momentum space. The same shall hold for linear co-ordinates; that is there is no preferred zero point in space (homogeneity in space and momentum space).

(c^1) The probability density of a particle of given p to be found in a range δq shall depend on the magnitude δq only, and not on the values q_1 and q_2 of which δq is the difference. The same for δp.

Although (c^1) is known from *classical* statistical mechanics as a separate theorem of a constant statistical density in qp-space, it is a consequence of (c), since dependence on q_1 etc., itself would indicate dependence on zero points. It will be noticed that the product $\delta q \cdot \delta p$, a range in qp-space, is invariant not only under Galileo but also under Lorentz transformations. The postulates (c) (c^1) merely confirm the homogeneity of space and momentum space. They are not *ad hoc* assumptions, and neither of them contains any hint of qp-periodicity. Yet, as will be seen, their combination with the general probability metric of interference leads, by mathematical necessity, to a function $\psi(q, p) = \exp(2i\pi qp/\text{const})$, identical with (1) when the constant is denoted by the familiar letter h. How large the constant h should be in terms of King Henry VII's arm from elbow to fingertip or in other conventional units, can of course not be answered *a priori*. It is significant, however, that h is small by ordinary standards because this safeguards an approximate validity of classical mechanics for the events of our daily lives.

23. *Origin of the wave function*

The proof that $\psi(q, p)$ must be of the complex periodic form (1) rests essentially on answering the third of the following questions:

1. Which function $\phi(p)$ satisfies $\phi(p)+\phi(p') = \phi(p \cdot p')$?
Answer: the logarithmic function, since $lg\, p+lg\, p'=lg(p{\cdot}p')$.

2. Which function $\chi(p)$ satisfies $\chi(p){\cdot}\chi(p')=\chi(p+p')$?
Answer: the exponential function, since $e^p{\cdot}e^{p'}=e^{p+p'}$.

3. Which function $\psi(p)$ satisfies $\psi(p)\psi^*(p')=\psi(p-p')$ where the asterisk indicates the complex conjugate?
Answer: the complex exponential function, e^{ip}, since $e^{ip}\, e^{-ip'}= e^{i(p-p)}$.

The detailed proof of (1) follows herewith. We begin with the general interference theorem of probabilities (ψ-matrix multiplication) applied to four orthogonal sets of states A, B, C, D:

$$(\psi_{AD}) = (\psi_{AB})(\psi_{BC})(\psi_{CD})$$

or written out in detail as a sum of products:

$$\psi(A_k, D_n) = \Sigma_j\, \Sigma_m\; \psi(A_k, B_j) \cdot \psi(B_j, C_m) \cdot \psi(C_m, D_n)$$

Any observable quantity T within the formalism of unitary transformation plays the part of a *tensor*, with ψ being the special tensor *unity*. That is the components of T satisfy a corresponding relation:

$$T(A_k, D_n) = \Sigma_j\, \Sigma_m\, \psi(A_k, B_j) \cdot T(B_j, C_m) \cdot \psi(C_m, D_n)$$

Let us now consider the special case that both sets of states B and C represent states of position, q and q', and that both A and D represent states of momentum, p and p'. The last formula then takes on the special form

$$T(p, p') = \Sigma_q\, \Sigma_{q'} \psi(p, q) \cdot T(q, q') \cdot \psi(q', p')$$

with summation over all values of q and of q'. Still more special, let T be an observable which depends on the coordinate q only, being defined as a function $T(q)$ so that

$$T(q, q') = T(q) \cdot \delta_{qq'} = T(q) \text{ for } q = q', \qquad (2)$$
$$= 0 \quad \text{for } q \neq q'$$

The last formula then reduces to a single sum on the right:

$$T(p, p') = \Sigma_q \psi(p, q)\, T(q)\, \psi(q, p')$$

Finally, using $\psi(q, p') = \psi^*(p', q)$, as well as the *constant* probability density in q-space for given p or p', the sum can be written as an integral† with constant weight factor in the integrand, that is

$$T(p, p') = \int T(q) \cdot \psi(p, q) \cdot \psi^*(p', q) \cdot dq \qquad (3)$$

We now make use of the requirement that, whatever is the special form of the function $T(q)$, the left-hand side of (3) shall depend on the *difference*, $p—p'$, only. This ought to be so also in the special case that $T(q)$ is chosen as a delta function with a sharp maximum at one place q, or at any other place q whatsoever. For a delta function T with a sharp maximum at q the last formula reduces to

$$T(p, p') = \psi(p, q)\, \psi^*(p', q)$$

where now both sides are to depend on $p—p'$ only. This can be so for any choice of q only when the function $\psi(p, q)$ is a complex exponential function

$$\psi(p, q) = \ldots \exp(\ldots ip \ldots) \qquad (4a)$$

according to the third problem at the beginning of this section. The dots indicate parts possibly depending on q but *not on* p.

When applying the same considerations, with q and p exchanged, to an observable $S(p)$ whose matrix element or transition value $S(q, q')$ is to depend on the *difference* $q—q'$ only, one arrives at the corresponding result that $\psi(q, p)$ must also be of the form

$$\psi(q, p) = \ldots \exp(\ldots iq \ldots) \qquad (4b)$$

with the dotted parts possibly depending on p but *not on* q.

† In my book, *Foundations of Quantum Theory*, of 1955, the summation was replaced by an integration without mentioning the implicit assumption that the density in q-space (and in p-space) for given p (and q) is *constant*. I take this occasion to correct this error of omission.

The last two equations together now leave only two alternatives:

$$\textit{Either } \psi(q, p) = \text{const} \cdot \exp(i\alpha q + i\beta p)$$
$$\textit{or } \psi(q, p) = \text{const} \cdot \exp(i\gamma qp)$$

where neither the constant in front nor α, β, γ depend on q or on p. Let us investigate the first alternative. When substituted in (3) it yields

$$T(p, p') = \int T(q)\, e^{i\beta(p-p')}\, dq$$
$$= e^{i\beta\,(p-p')} \int T(q)\, dq = \text{const}\; e^{i\beta(p-p')}$$

which would mean that the transition value $T(p, p')$ is independent of the form of the function $T(q)$, that it is the same for all observables $T(q)$ whatsoever, which of course does not make sense. Only the second alternative remains. It reads, when replacing the letter γ by $2\pi/h$:

$$\psi(q, p) = \text{const} \cdot \exp(2i\pi qp/h) \tag{5}$$

q.e.d. Quantum dynamics is thus reduced to:

(*a*) the dependence of 'transition values' of observables on p- and q-intervals only;

(*b*) the constant statistical density in q- and in p-space for given p- and q-values respectively;

(*c*) the general probability metric of unitary transformation, that is the interference law for probability amplitudes.†

For later use substitute (5) into (3) and obtain

$$T(p, p') = \int T(q) \cdot \exp[2i\pi(p-p')q/h] \cdot dq \tag{6}$$

It is well known that the wave-like function $\psi(q, p)$ is equivalent to the Schrödinger rule, and to the Born commutation rule, $qp - pq = h/2i\pi$. The latter somewhat enigmatic-looking rule is a special case of the non-commutability of two observables u and v which are incompatible with one another, so that there is no 'state' in which u and v have definite values simultaneously. The product quantity uv simply

† If different particles had different constants h they could never be in quantal interaction.

is not an 'observable', neither is vu. Niels Bohr(2) writes in this connection: 'In the quantum formalism, the quantities by which the state of a physical system is ordinarily defined, are replaced by symbolic operators subjected to a non-commutative algorithm involving Planck's constant.' This sounds very enigmatic indeed. But what does the algorithm of non-commutability actually imply? It tells us that in the formalism of unitary transformation, when u and v are two observables = tensors and when k and j refer to two well-defined states:

$$(uv)_{kj} \neq (vu)_{kj}, \text{ yet } (uv)_{kj} = (vu)^*_{jk}, \text{ hence } \\ |(uv)_{kj}|^2 = |(vu)_{jk}|^2 \tag{7}$$

In contrast, commutability of u and v would mean

$$\left.\begin{array}{l}(uv)_{kj} = (vu)_{kj}, \text{ hence} \\ |(uv)_{kj}|^2 = |(vu)_{kj}|^2\end{array}\right\} \text{(wrong!)} \tag{8}$$

But which is more self-consistent, (7) or (8)? Certainly (7), that is non-commutability. Indeed, on both sides of (7) u is associated with k and v with j (whatever the physical meaning of the word 'associated' may be). In (8), however, the 'associations' are different on the two sides of the equation, which is inconsistent from a purely formal point of view, hence would lead to an inconsistency in any physical model, too. Non-commutability (8) is just what one would have to expect on grounds of simple postulates of self-consistency, formal and physical.

24. *One-dimensional crystal*

Let us discuss a few simple consequences of the periodicity of ψ in q and p. Consider a body which is periodic in the q-direction with periodicity l so that all observables $T(q)$ have the same values at q as at $q+l$ and $q+2l$, etc. In this case $T(q)$ may be expanded as a Fourier series:

$$T(q) = \Sigma_n\, T_n \cos\left(\frac{2\pi nq}{l} + a_n\right) \tag{9}$$

$$= \Sigma_n\, \tfrac{1}{2}T_n \cdot \exp\left(\pm\frac{2i\pi nq}{l} \pm ia_n\right)$$

where the $\pm$ indicates a sum of two terms, one with $+$ and the other with $-$ sign. A body with observables of this form may be denoted as a one-dimensional crystal, schematically as a set of parallel planes perpendicular to the q-direction and at distance l from one another (Fig. 5), filled with matter according to a density pattern which repeats itself periodically.

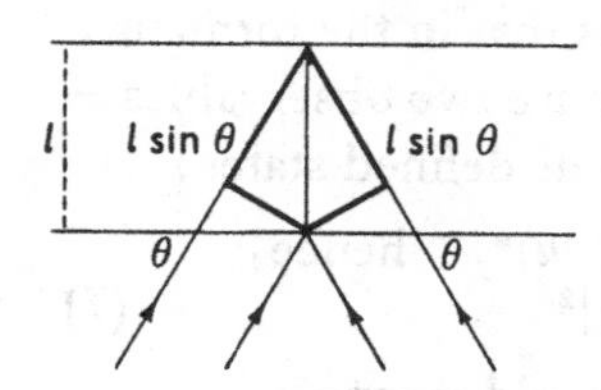

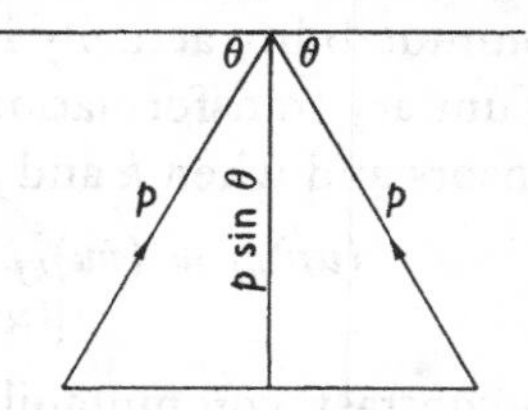

Fig. 5. Duane's elaboration of Bragg's principle of selected reflection from a set of parallel planes.

In order to investigate the qualities in momentum space of this 'crystal', substitute (9) into (6):

$$T(p, p') = \Sigma_n \tfrac{1}{2} T_n \int \exp\left[2i\pi\left(\frac{p-p'}{h} \pm \frac{n}{l}\right) \cdot q \pm i a_n\right] dq \quad (10)$$

The integrand is periodic in q so that the integral consists of positive and negative contributions which cancel one another unless the factor of q in the exponent vanishes. That is, $T(p, p')$ has non-vanishing values only under the condition

$$p - p' = \pm nh/l \quad (11)$$

For p-differences not satisfying this condition the 'transition values' $T(p, p')$ vanish; they exist only for $\Delta p = \pm nh/l$. The same may be expressed in these words: Our one-dimensional crystal is, from the dynamical point of view, a body which can change its momentum in the q-direction only in quantized amounts $\Delta p = nh/l$, that is in multiples of the basic quantity h/l where l is the periodicity along the q-axis.

It may be noticed that only *p-increases* and *decreases* of a tested object can be determined experimentally by measuring p-decreases and increases of another object in reaction to the tested object. The 'theory of measurement' in its full scope enters here. The two descriptions, of $T(q)$ in space as a function of periodicity l, and of $T(p, p')$ in momentum space signifying quantized momentum changes $\Delta p = nh/l$, are *complementary* descriptions of the same 'crystal'.

The first to notice that a crystal of parallel planes can *also* be described as a body with quantized momentum reactions was W. Duane(3); he thereby gave a significant contribution to the development of modern conceptions in quantum theory. Duane remarked that the wave condition of selected reflection from a set of parallel planes of distance l,

$$2l \sin \theta = n\lambda \text{ (Bragg)} \tag{12}$$

as the result of interference of waves λ reflected from the planes, may be multiplied by the constant h and then reads

$$2\frac{h}{\lambda} \sin \theta = n \cdot \frac{h}{l}, \quad \text{or} \quad 2p \sin \theta = n\frac{h}{l} \text{ (Duane)} \tag{13}$$

The left-hand side describes the momentum change of incident particles perpendicular to the planes. The right-hand side must therefore represent the momentum gain by the crystal perpendicular to the planes, in agreement with (11).

25. *Three-dimensional crystal*

A three-dimensional crystal is composed of basic cells characterized by three ground vectors a_1 a_2 a_3 filled with matter according to a density pattern repeated in every cell. The direction of a set of parallel lattice planes is defined by a plane laid through the endpoints of three vectors n_1a_1, n_2a_2, n_3a_3 with integers n_1 n_2 n_3 which also determine the distance l between subsequent planes. The Bragg wave condition (12) again determines the angles of selected reflection according to wave interference theory.

The same selected angles can be obtained also from a construction in the 'reciprocal lattice', the basic cell of which is formed by three ground vectors b_1 b_2 b_3 which are perpendicular to the walls of the former space lattice, and whose magnitudes are reciprocal to the perpendicular distances $d_1d_2d_3$ between the walls of the space lattice cell

$$b_1 = 1/d_1, b_2 = 1/d_2, b_3 = 1/d_3$$

A point of the reciprocal lattice is reached from its zero-point by a vector

$$R = k_1b_1 + k_2b_2 + k_3b_3$$

with three integral numbers $k_1k_2k_3$. The reciprocal lattice is presented by Fig. 6 with O as zero-point, which may be chosen at will.

The following construction for the selected reflection of incident waves λ has been given by P. P. Ewald.(4). Draw a vector MO of length $1/\lambda$ in the direction of the incident waves ending in a point O of the reciprocal lattice. Then draw a sphere around M with radius OM. If this sphere runs through, or very close to, another point of the reciprocal lattice, such as the point H, draw the vector MH; it will represent a reflected direction, when MO is the incident direction. Geometrical considerations show that this construction is equivalent to that of Bragg.

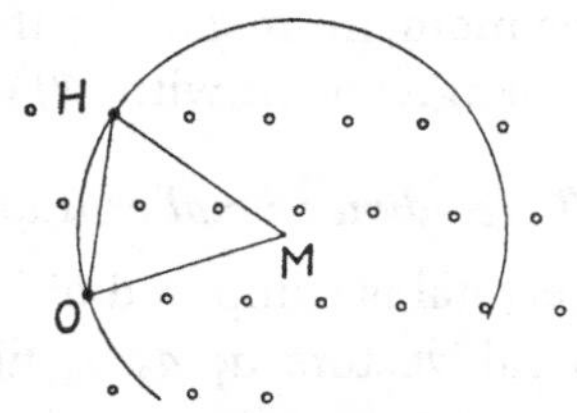

Fig. 6. Ewald's construction for selected reflection.

Ewald's construction may be interpreted as a construction n momentum space when multiplying all lengths in the diagram of Fig. 6 by the factor h. The vector MO then represents the momentum of the incident particle of magnitude h/λ, whereas MH is its momentum vector after reacting with the crystal. The magnitude of the particle momentum remains unchanged during the reaction with the crystal; only its direction is changed according to the momentum vector equation $MO + OH = MH$. Thus OH represents the momentum vector gained by the particle. A crystal can thus be described as a mechanical system which is capable of giving out (or receiving) only such directed momenta as are represented by connections between any two points in the reciprocal lattice, multiplied by h. The two descriptions of the crystal, as a periodic arrangement of matter in space, and as a mechanical system giving out and receiving quantized momentum vectors, are complementary. However, according to the unitary particle theory of matter, only the second description is a realistic report on the diffraction of *matter*; the wave description deals with probability amplitudes of matter particles, not with real waves. We cannot discuss photons *versus* light waves since this would lead into details of the quantum theory of fields. Refer to the end of section 33.

26. *Bohr frequency and Heisenberg uncertainty*

It is known from ordinary mechanics that energy E and time t play the part of conjugate observables similar to p and q. One only has to substitute E for p and t for q in the above developments in order to arrive at new consequences of our ground postulates (section 22) concerning conjugates. The first conclusion is that the amplitude function $\psi(E, t)$ must have the form

$$\psi(E, t) = \text{const} \cdot \exp(2i\pi Et/h) \tag{14}$$

Suppose, further, that those observables of a certain object which are functions of t have the simple harmonic form

$$T(t) = T_o \cdot \cos(2\pi\nu t + \alpha)$$

The object would then be called a 'harmonic oscillator' of frequency ν. Another object (atom) may have observables $T(t)$ of the more involved form

$$T(t) = \Sigma_n T_n \cos(2\pi\nu_n t + \alpha_n) \tag{15}$$

Substitution into the integral:

$$T(E, E') = \int\psi(E, t)\ T(t)\ \psi(t, E')\ dt$$

then yields the result, similar to (11), that $T(E, E')$ has non-vanishing values only for

$$E - E' = \pm h\nu_n \text{ or } \nu_n = \frac{|E - E'|}{h} \tag{16}$$

This is Bohr's frequency condition. Thus, when observing the emission or absorption spectrum $\nu_1\ \nu_2\ \nu_3$ . . of our object (atom), we can infer from it a set of energy levels characteristic of the object, which yield the observed frequencies according to the Bohr frequency condition. (5) and (14) lead to the quantum prescriptions of Born and Schrödinger, to Heisenberg's uncertainty relation, to Bohr's complementarity, and all the rest. If these matters were once beyond our understanding, if they were 'riddles', then they have now been solved by reduction to almost obvious non-quantal postulates.

A short discussion of the famous Heisenberg uncertainty relations may be added here. They are obtained by translation into the language of particles of the fundamental rule for *resolving* two waves, one of frequency ν and wave number $\tilde{\nu}$ ($= 1/\lambda$), the other of frequency $\nu + \Delta\nu$ and of wave number $\tilde{\nu} + \Delta\tilde{\nu}$, namely, the rules

$$\Delta\nu \cdot \Delta t \sim 1 \text{ and } \Delta\tilde{\nu} \cdot \Delta q \sim 1 \tag{17}$$

where Δt is the time interval and Δq is the space range over which the two waves are permitted to be observed. Equation (17) tells us that the frequency ν of a wave train can be measured only with a margin of uncertainty $\Delta\nu \sim 1/\Delta t$ during a time allowance is Δt. Similarly, the wave number $\tilde{\nu}$ can be found only with a margin of uncertainty $\Delta\tilde{\nu} \sim 1/\Delta q$ when the space range is allowed for the observation Δq.

Multiplication of (17) by the factor h and substitution of $E = h\nu$ and $p = h\tilde{\nu}$, yields the Heisenberg relations

$$\Delta E \cdot \Delta t \sim h \text{ and } \Delta p \cdot \Delta q \sim h \tag{18}$$

Their meaning is: when a particle is permitted to be observed only during a time interval Δt, its energy remains uncontrollable with margin $\Delta E \approx h/\Delta t$. And when a particle is permitted to be observed only within a space range Δq, its momentum is uncontrollable up to a margin $\Delta p \approx h/\Delta q$. The Heisenberg uncertainty relations for particles can also be derived directly from the quantum mechanics of particles, and chiefly from the fact that the probability amplitudes $\psi(q, p)$

and $\psi(E, t)$ are of the exponential form $\exp(2i\pi qp/h)$ and $\exp(2i\pi Et/h)$ respectively.

I cannot follow the current interpretation of Heisenberg's uncertainty relation, namely that

(1) A simultaneous exact x- and p_x-value of a particle can never be ascertained experimentally beyond the product margin h.

(2) Therefore one should not even imagine that a particle could *have* an exact x- and p_x-value at the same time (Bohr).

(3) Hence the classical *concept* of an ordinary particle must be abandoned—which is supposed to be of great philosophical importance.

However, (2) and (3) stand and fall with (1), and (1) is hardly tenable; it rests on a misinterpretation of Heisenberg's relation. Indeed, as Karl Popper pointed out twenty-five years ago, and as H. Margenau has re-emphasized recently: when the position x of a particle is ascertained with margin Δx, for example just after passing through a slit, then its simultaneous momentum p_x can be ascertained, on the grounds of its trajectory to a point of impact, with an accuracy far surpassing Heisenberg's Δp_x. However, many (almost) exactly observed p_x-values emerging from Δx display an uncontrollable statistical dispersion $\Delta p_x \sim h/\Delta x$. In other words, if there is to be an observed spread Δp_x, it can only be the spread of more exactly determined p_x-values. All this does not impair Heisenberg's formula, but it concerns its physical interpretation.

It was seen above that the uncertainty relations for particles can be obtained by translation of the wave laws for the resolving power. Quite generally, one often succeeds in obtaining relations between particle events by the method of first translating E and p into wave terms, ν and $\tilde{\nu}$, then applying wave theory, and finally retranslating ν and $\tilde{\nu}$ into mechanical E and p of particles. The fact that this method succeeds, that wave theory closely corresponds to the probability theory for particles, may be *described* by the term 'duality'. It is *explained*, however, by reduction of the wave-like quantum rules to simple non-quantal postulates, so that one does not need specific *ad hoc* quantum postulates in their own right.

27. *The diffraction experiment*

This book deliberately refrains from dealing with light waves versus photons, since light belongs to the relativistic theory of fields, in spite of occasional similarity with matter. We are concerned here primarily with the elementary quantum mechanics of *matter*, although the same formal rules also apply to the harmonic components of fields.

A great difficulty for the unitary particle interpretation has often been seen in the phenomena of diffraction, with maxima and minima of intensity observed on a film C irradiated by a matter ray from a source A, passing through a screen B with two (or more) parallel slits B_1 and B_2. The pattern looks indeed as though it were due to the interference of waves of length $\lambda = h/p$, where p is the corresponding particle momentum. From the corpuscular viewpoint one may ask: how can those particles which pass through B_1 'know' of the presence of the slit B_2 through which they do not pass, and yet preferably travel towards those places on the film C where wave theory predicts an intensity maximum produced by interference through both slits simultaneously. Before answering this question, notice first that quite generally the phenomenon of 'statistical co-operation' of mutually independent particles in producing any statistical average whatsoever is a miracle indeed which cannot be causally explained (Chapter I). However, when statistical co-operation is once taken for granted, as it must be taken in *any* statistical theory, then it is not quite so enigmatic any more that this co-operation has the form of an *interference* of probabilities as was seen in Chapter III. In the case of diffraction through two parallel slits yielding an interference pattern, remember that the same interference pattern would result also from *reflection* of the matter ray from a polished screen with two non-reflecting lines B_1 and B_2. The former question 'How can a particle passing through B_1 "know" of the existence of B_2 through which it does not pass?' here reads 'How can a particle *not* reflected from line B_1 "know" of the existence of the other line B_2 from which it is *not* not-reflected?' If the latter question is recognized as spurious, the same holds for the former; apparently something is taken for granted here which ought not be taken for granted.

Indeed, in order to understand the interference pattern from the particle point of view, one must consider a mechanical interaction between the incident particle and the screen *B as a whole.* How to do this in detail can best be learned from a consistent translation of the theory of wave *appearances* into particle realities. We here follow the historical development which also happens to be the most systematic one. Instead of treating of a screen with only two slits or non-reflecting lines, first consider a periodic arrangement of matter, that is a *crystal,* and the diffraction of particles through it. According to Duane (section 24) the same crystal which from the wave point of view offers a periodic arrangement of lattice planes at distances l in space, acts as a mechanical system capable of changing its momentum perpendicular to the lattice planes only in quantized amounts of magnitude

$$\frac{h}{l}, \frac{2h}{l}, \ldots \frac{nh}{l}, \ldots$$

Hence it is misleading to maintain that every electron acts as though it were spread out in space. It rather is the crystal or the screen with slits which is spread out in space and acts as a mechanical unit to the incident particles.

Duane's important demonstration that Bragg's selected reflection law can be understood on the basis of mechanically ruled processes for particle interaction with the crystal as a whole, was soon followed by the Ehrenfest and Epstein([5]) particle theory of diffraction through any number of parallel slits or lines on a screen; the screen as a whole gives out selected momenta corresponding to the periodicities of its space structure, again revealing the spuriousness of the question 'How can a particle know, etc.'. It does not need to know since it reacts with the screen and its slits or lines as an entire mechanical system.

Schrödinger on the contrary strove for a pure wave interpretation of various atomic phenomena; he found(6) that the Compton effect, which had previously been derived from particle theory with conservation of energy and momentum, can also be understood as an interference effect of waves. On the other hand, he showed(7) that the Doppler effect, which had always been regarded as a convincing proof of the wave nature of light, can also be understood in terms of photonic particles emitted from a moving source. All this demonstrated a formal *equivalence* of the two modes of description as far as overall statistical results are concerned. Schrödinger regarded the wave theory as describing the *real* physical situation without being able, however, to explain individual discrete corpuscular phenomena. Born's statistical particle theory is to be preferred as the *real* thing since it not only accounts for discrete events but at the same time admits of an explanation for wave-like probability interference in general (Chapter III) and for the periodic wave-like relation between co-ordinates and momenta (Chapter IV) which is the backbone of quantum dynamics.

28. *Summary*

(*a*) Let q and p be variable between $+\infty$ and $-\infty$. Let the components $T_{qq'}$ and $T_{pp'}$ of a tensor T in the space of unitary transformation depend only on the differences $q-q'$ and $p-p'$ respectively. Let integrals over q at fixed p, and over p at fixed q, be carried out with *constant* weight factors. This leads to the conclusion that the unit tensor ψ must have 'components' $\psi_{qp} = \psi(q, p) = \exp(2i\pi qp/\text{const})$.

(*b*) The quantities q and p are identified as co-ordinates and canonically conjugate momentum components. The constant in the exponential function is Planck's action constant h, so that $\psi(q, p)$ is a 'wave function' with wave length $\lambda = h/p$.

(*c*) The foregoing definition of conjugacy agrees with classical dynamics. There, too, only q-differences and p-differences, that is invariance with respect to parallel displacement in q- and p-space is of physical significance; and classical statistical mechanics, too, rests on a constant probability density in q- and p-space. The quantum periodicity is thus explained as a consequence of familiar features of q and p in classical dynamics, namely *invariance* with respect to Galileo and/or Lorentz transformation, combined with the general probability metric of Chapter III.

CHAPTER V

QUANTUM FACT AND FICTION

'Query 64: Whether mathematicians who are so delicate in religious points, are strictly scrupulous in their own science? Whether they do not submit to authority, take things upon trust, and believe points inconceivable? Whether they have not their mysteries, and what is more, their repugnances and contradictions?'

BISHOP GEORGE BERKELEY

When speaking of fact and fiction in quantum theory, I know of course that the experts handle their actual problems in atomic physics with the greatest dispatch and virtuosity. Yet when listening to their words, to the meaning they attribute to the formalism, to the quantum ideology and its alleged significance within the theory of knowledge, I cannot help wondering whether a clearer picture might not be attained after the elimination of a few generally accepted notions which do not fit into our present state of knowledge any more. During the 1920's when Heisenberg-Born-Jordan's and de Broglie-Schrödinger's ingenuity opened a new path towards solving atomic problems, physicists had reason enough to be startled by the abstractness of their method. At that time they felt compelled to *talk* their way out of antinomies and paradoxes rather than solving them by the methods of theoretical physics. Misconceptions adopted during the creative stage of quantum mechanics have shown a remarkable staying power, however, and together with a loose terminology are now canonized as the only true faith, in spite of semantic tricks (see below) which would be condemned in any less elusive scientific discipline. The mathematical sign language with its complex symbols and non-commutative matrix algebra has become a veil shielding the simple meaning of the quantum laws from the scrutiny of common sense, culminating in Eddington's assertion that 'we have learned that the exploration of the external world by the methods of physical science leads not to a concrete reality but to a shadow world of symbols . . . as one of the most significant recent advances'. If so it would not be an advance but a complete reversal of natural science whose object was and still is: to take a real world for granted, then explore and describe it in terms of objective laws *not* of our own making. If in atomic dimensions some observations seem to favour particles, and others seem to favour continuous waves as the real constituents of matter, then it is pure evasion to regard our temporary indecision in this dilemma as being on a level with the deep *philosophical* problem of physical reality which has troubled thinkers from antiquity to modern times.

Eddington's view that 'physical science leads not to an objective reality' is akin to that of theorists, who hold that quantum theory does not deal with objective data collected in statistical ensembles of experiments, but with subjective states of mind so that the time of the 'transformation from the possible to the actual' can be shifted forward and backward at our discretion (refer to sections 30 to 32). Similar to Eddington's shadow world of symbols, Bohr's and Heisenberg's dualism evades the question of the *real* constitution of matter by insisting on two opposite though complementary 'pictures'. (As seen before, particles are real things, ψ-waves are not.) Having given a name to the dualistic antinomy, the entrenched dualists thereafter simply deny that there is a problem left any more. When further pressed for an answer as to the real constitution of matter, they ask: 'What do you mean by "real" particles and by "real" waves anyway, and what is the use of "explanation"?' My answer is that a 'real thing' is characterized by constantly recurring qualities. Thus, an electron with its discrete charge and rest mass is a real thing. But the probability curve or function describing past statistical experiences is not a real thing, even when the curve looks wave-like. Also, using the good Doctor Johnson's language: You can kick a stone, and you also can kick an electron, and even a water wave and an electromagnetic wave, and be hurt by them, proving their 'reality'. But you cannot kick, or be hurt by, a curve representing probabilities of possible events. And 'explanation' means reduction of strange and complicated special situations to familiar and simple general principles.

Actually, no philosophical profundity is needed for the understanding of the quantum theory if one holds consistently to the *unitary* statistical particle interpretation of Max Born, which most physicists adopt in their deeds, though not always in their words. I have tried in what follows to point out misleading notions produced by *duality*, *la grande illusion*, so that we may return from symbolism and subjective pictures to plain physics again without philosophical ornaments. If my *critique* deflates current modes of quantum talk, I ask for the indulgence of the reader who knows that 'once allegiance is given to any school, a complete change of intellectual habits is required before we can even become aware of other schools. And then our first reaction is one of pain and distaste' (G. Sykes, *Harper's Magazine*, March 1958). If Bohr and Heisenberg come under particularly sharp scrutiny, it will hardly be necessary to add that my admiration for their achievements in theoretical physics is as great as ever.

29. *Duality and doublethink*

The discovery of matter ray diffraction in the 1920's seemed to involve a serious paradox concerning the true constitution of matter, similar to the wave-versus-photon antinomy in the constitution of light revealed earlier by Planck and Einstein. It was most comforting at that time to listen to the positivistic siren song that one should not insist on any 'true' constitution of matter at all, that both particles and waves are but convenient 'pictures', that neither of them has a claim to exclusive reality. Moreover, it was most gratifying to learn from Niels Bohr that the two pictures never conflict with one another in the interpretation of individual observations, because those experiments which seem to reveal a corpuscular exchange of energy and momentum display a fuzziness in space and time, and *vice versa*, a fact inherent in Heisenberg's uncertainty principle. Wave-particle duality, bridged by mutual complementarity, might thus be regarded as a fundamental feature of matter and light which cannot be reduced to anything more elementary; it must be accepted at face value without further explanation. The most perfect expression of this basic duality is found in Heisenberg's Chicago University lectures([1]) of 1930. However, as much as I admired Heisenberg's methodical presentation at the time, I cannot believe any more in his message that wave-particle duality is fundamental.

To my mind, the resignation to an inscrutable duality and complementarity may have been reassuring to an earlier generation in despair over an apparent antinomy. But even the most persuasive arguments that duality is the ultimate aspect of the microcosm cannot keep all physicists at all times from asking questions as to how and why an electron can alternately appear as a discrete particle and as a continuous wave field (quoting von Weizsäcker, see below). As to the question whether one might, by further analysis, arrive at a clear decision about a *real* constitution of matter after all, either real waves with corpuscular appearances or the other way around, one will remember Schrödinger's attempt of interpreting particles as high crests of real substantial waves, and the disappointment when this *unitary wave theory* failed, because high wave crests flatten out rapidly and fuse into one another. Born's counter proposal, his statistical particle view, proved of enduring value. However, instead of pursuing and perfecting the *unitary particle interpretation* consistently, it is the present fashion to accept the statistical particle interpretation only on workdays and cling to the dualistic doctrine of waves and particles on Sundays.

This attitude violates one of the principal rules of orderly thinking, namely the rule: do not indulge in false opposites. Do not construe an antithesis between a thing characterized by definite invariant quantities on the one hand, and *one* of its many variable attributes or qualities on the other. Do not contrast a snake as a thing to its occasional wavy shape in order to defend a duality of snake substance. Quantum theorists, however, constantly violate this rule when they proclaim an opposition between particles as the substance of matter on the one hand, and the occasional wave-like statistical distribution of same particles on the other. It may seem undignified to draw a parallel between a crude snake duality and the deepest and most delicate tenet of modern quantum theory. Yet the term 'duality' should *never* be applied to contrasting a thing (particles) to one of its many qualities (statistical disposition of particles). Disregarding this rule has led to many misconceptions in quantum mechanics. After Born's statistical particle interpretation is once accepted, 'duality' becomes an anachronism, bound to give a distorted view of the actually unitary character of the quantum theory of particles including their wave appearances, which may have been demonstrated in the previous chapters. If the experts cling to their accustomed language and still use the notion of duality for heuristic purposes, there is no harm. But the constant harping on an antinomy which has long been removed by Born, is not conducive to clarity, as a glance at the popular literature will show.† Textbooks, if they do not simply shun the issue, and public addresses of authorities are hardly better.

Sir James Jeans proclaims in his Presidential Address(2) to the British Association of 1934: The particle picture is a materialistic picture which caters for those who wish to see their universe mapped out as matter existing in space and time; the wave picture is a determinist picture which caters

† As an extreme case of public enlightenment we quote the following contribution to space travel from an American Press report (A.P., July 1955): 'Scientists today know that matter is composed of vibrations . . . therefore, the human body, which is matter, can be broken up into vibrations and relayed electronically anywhere in the world. . . . I believe that before space travel is developed, someone will come up with a way to devibrate the human body, relay it by wires and waves, and revibrate it at the point of destination.' The devibrating will not prove too hard a task, but the revibrating may encounter unexpected difficulties. [*Author.*]

for those who ask the question: 'what is going to happen next?" . . . The wave picture which observation confirms in every known experiment, exhibits a complete determinism; (but it is) one of waves, and so, in the last resort, of knowledge.' As to energy conservation 'waves of knowledge are not likely to own allegiance to this law'. I do not blame Jeans for echoing the general confusion prevalent in 1934. But even twenty years later, when the dust should have settled, von Weizsäcker([3]), an official protagonist of the Copenhagen School, in his book, *The World View of Physics*, asks: 'In what respect does quantum mechanics differ from classical physics? It has discovered that the same physical object, for example an electron, appears in two different seemingly mutually exclusive forms . . . particle or wave . . . cloud chamber photographs *versus* interference. . . . Now, what meaning can there be in the statement that an electron is both particle and field?' The logical answer to this query seems to be: 'No meaning at all,' since an electron is a particle but the statistical behaviour of many electrons is not what an electron *is*. Von Weizsäcker himself, however, avoids answering the question by describing the incompatibility of *particle* position and *particle* momentum, thereby applying the term 'wave parcel' in the interpretation as a 'probability parcel' for *particles*, thus all the time dealing with *particles* and their distribution in space and time, yet suggesting that this pure particle theory supports particles and waves on equal terms. Later he writes: 'The same atom behaves in some experiments like a spatially concentrated particle, in others like filling the whole space.' This sounds rather alarming. Actually, atoms not only *are* particles by virtue of their invariant masses, charges, etc., but behave exactly as particles ought to behave statistically. For although the probability correlation law *reminds* us of wave interference, it is a necessity for particles under the supposition that there is a general probability correlation law at all (Chapter III). Of course, von Weizsäcker does not believe in material waves. But then, what does he mean by his atoms 'filling the whole space'?

Another anachronistic plea for duality, to quote but one among innumerable defenders of the faith, is made by R. Havemann([4]): 'There have been differences of opinion among the physicists as to which of the two pictures is the true one. Some have declared the corpuscular picture as true, with waves as "probability waves". Others, in particular Schrödinger, consider the wave picture as true. Today the majority of physicists, correctly in my opinion, take the standpoint that both pictures are equally true and equally untrue, in so far as none of them alone is sufficient, and that one picture must rather be complemented by the other.' I agree with the majority in rejecting the wave interpretation. But accepting Born's particle interpretation and yet insisting that neither picture is true, is mixing up the deep philosophical problem of reality with a technicality of atomic physics. When Havemann later clearly concedes: 'The observed actual phenomena in micro-physics are all corpuscular,' he is right, even with respect to the probability correlation law which looks like wave-interference. Yet in the same breath he becomes impartial again and adds: 'But the potentialities for particles can be represented by pure wave mechanics.' The word 'but' is decisive here; it is to signify that there is a paradox. I would like to replace the 'but' by 'hence', and add: 'hence there is no duality paradox, except as as a superficial appearance'. Strangely enough, Max Born himself makes belated concessions to duality. In a lucid article([5]) on 'Physical Reality' in 1953 he equates the popular concept of *real things* with the more precise concept of *invariance* of 'Gestalt' in everyday life, and of invariant constants in physics. 'Invariants are the concepts of which science speaks in the same way as ordinary language speaks of "things" and which it provides with names as though they were ordinary things.' Electrons are real things because they have invariant rest masses, charges, and spins, even if they are said never to have simultaneous positions and momenta. Similarly, water waves and electromagnetic waves are real things in the usual sense of the word. But then Born continues: 'Why then should we withhold the epithet "real" even if the waves represent in quantum theory only a distribution of probability?' My answer is: 'For the same reason that sick persons are real things; whereas the "curve" which indicates the maximum probability for a person to be sick, that is the peak of an epidemic, is not a real thing. Regarding a wave-like disease curve as a thing of equal "reality" as a sick person may be a harmless play with words since everybody recognizes its deviousness. In quantum theory, however, the declaration 'why then should we withhold the epithet', etc., gives the impression that some elusive yet profound issue is at stake which must be protected at all costs from profane comparison with ordinary situations.

It may seem pedantic to criticize illogical habits of speech which have not hampered the successful application of the quantum method to actual problems in atomic physics. The trouble is, however, *first* that the student presented with this kind of doublethink has to pass through a long period of bewilderment, until he finally learns to grope his way more or less instinctively through the 'quantum mess'. And *second*, the same illogical talk has obscured the problem as to *why* particles should ever behave in a wave-like fashion; even the question is considered by the dualists as naïve, or as 'irrational'.

Theorists defending particles and waves on an equal footing often refer to an investigation by Klein, Jordan and Wigner where statistical particle mechanics is translated formally into an equivalent wave theory in three-dimensional space. However, equivalence with respect to observation is heavily outweighed by non-equivalence in every theoretical aspect. When comparing the K-J-W formalism with the usual particle theory one finds *deviousness* rather than directness and economy (nobody will solve atomic problems by the roundabout continuum method), *complexity* rather than mathematical simplicity (few physicists even know how to handle the K-J-W method), and *ad hoc character* rather than explainability (the complicated qualities of K-J-W's fluid would be hard to justify from the standpoint of a fluid; they must be introduced for the purpose of equivalence). Mathematical transformations of a theory always safeguard observational equivalence. But what counts for accepting a theory as adequate or true, is directness, simplicity, and lack of *ad hoc* elements. (Franklin's theory of lightning is more true than Homer's.) And on this account the particle theory of matter is *not* equivalent to K-J-W's theory.

So far I have dealt with the problem of duality only in the non-relativistic quantum theory of matter where particles are 'real', and ψ-waves merely signify certain statistical features of the particles rather than material waves. The objection has been raised that, when leaving the narrow non-relativistic domain, the dualistic contrast between waves and particles appears again, and the 'principle' of duality is indispensable. For example, it may be asked: which is 'real' and which is 'mere appearance', photons or light waves? To this my reply is first: a photon is an obsolete concept, still of heuristic value, but abandoned in field theory. Quantum mechanics divides the field into 'field oscillators' or other quasi-mechanical constituents with generalized co-ordinates q and momenta p, which are subjected to exactly the same formal rules as the q's and p's of matter particles. They also give rise to ψ-functions and pq-periodicity—for *reasons* explained in Chapters II and III, without the necessity of having *faith* in a 'principle' of wave-particle duality.

The field oscillators or any other 'components' of the field are certainly 'real things', whereas the ψ-functions or ψ-waves in the space of the field co-ordinates are not. Of course the reality of the field oscillators is not as individualized as that of matter particles with their invariant rest masses, charges, spins, etc. On the other hand, matter particles can be annihilated or created 'out of nothing'; but quantum theory treats such phenomena simply as transitions from a positive energy level to a negative 'hole' or *vice versa*, and the formal quantum rules are always the same. The big question today is that of the variety of Hamiltonian energy functions which define the particles and fields, into which the formal quantum rules are fed. But there is no need for appealing to a 'principle' of duality of pre-Born days.

30. *Complementarity, subjective, interpretation*

The prolonged effort to save dualism by denying that neither 'picture' is true, has resulted in an almost total eclipse of the intellectual curiosity to find out *why* particles produce wave-like statistical appearances. Instead of laying bare the deeper correlations which lead to dualistic appearances, the primary goal has been that of defending and then enlarging the doctrine of duality and complementarity. Niels Bohr in particular believes to have found confirmations of complementarity not only in atomic physics but in other fields of human knowledge([6]), in biology, psychology, sociology, and so forth. 'We may truly say that different human cultures are complementary to each other.' Inspired by this vista of universality, one of Bohr's followers, J. A. Wheeler, exclaims: 'Complementarity represents, in one sense, the most revolutionary conception of our day.' And if the Master himself offers 'justice versus charity' as a further illustration, then we cannot be surprised if others wish to participate in this 'most revolutionary conception', for example when Jean-Paul Sartre([7]) discovers complementarity also in the eternal contrast of male and female (without saying which is waves and which is particles). Here one is reminded of K. R. Popper's poignant remarks([8]) on the all-embracing theories of Freud and Adler: 'It was precisely this fact that they always fitted, that they were always confirmed—which in the eyes of their admirers constituted the strongest argument in favour of these theories. It began to dawn on me that this apparent strength was in fact their greatest weakness.'

Margenau([9]) has posed the question whether in quantum theory 'physics has seriously begun to describe human knowledge, a subjective aspect of the mind, in terms of differential equations involving physical constants. The point I wish to make is that we are not forced to this conclusion'. I emphatically agree with Margenau that quantum theory deals with objective situations, and not with packets of expectation spreading out in space and suddenly collapsing as though through a kind of telepathy. The quantum theorist, as every scientist, co-ordinates data recorded by (macrophysical) instruments, data which are 'objective', that is *reproducible* by immediate repetition of the same test with the same object, although it may often be difficult or impossible to recognize 'the same atom' in the 'same state' by means of the 'same instrument'.

It is true that the outcome, the passing or not passing of a particle in an individual test, is not uniquely determined by its previous state but is ruled statistically. Still, subjectively tainted expressions such as uncertainty, probability, expectation, etc., signify statistical ratios of *objective data* recorded in many actual experiments, collected in tables or represented as functions. The fact that the physicist takes past statistical results as a topic for his personal reflection, calculation, expectation, and as betting odds for future tests, etc., is unessential; it is not even characteristic of quantum theory. Any classical law, such as $s=\frac{1}{2}at^2$, is first a generalized condensation of past experience; it also can be used for planning future experiments. Whether the odds of expectation are spread statistically over a wide margin, or condensed in one value with certainty, is a matter of degree, though of vital importance to the physicist. Still, quantum theory, like classical theory, connects data recorded by instruments. Speaking of waves of expectation in space, or in an observer's brain, is as tortuous as speaking of six threads of expectation in a dice game which suddenly break when a die is cast, or is seen by an observer to be cast. Notions like this are left-overs from a period when one believed in ψ-functions as describing physical states of vibration. Today we have the even worse misinterpretation of ψ-functions as describing states of knowledge, or lack of knowledge. Insurance companies have permanent tables for statistical odds under various physical conditions; similarly, quantum theorists have P- and ψ-tables and functions. But why such a table or function should suddenly collapse or contract (Heisenberg-von Neumann 'picture', section 33) upon an actual event is beyond the writer's capacity to grasp, in quantum physics as well as in insurance.

One of the foremost achievements of quantum physics has been the discovery that co-ordinates q and momenta p are incompatible, that there are no qp-combination states. This is certainly of great physical interest—as interesting as the opposite result that energy E, angular momentum A, and z-component of A are compatible, that is that there are reproducible EAZ-combination states. But why is it of great *philosophical* significance that some groups of observables are compatible, and others are not? Flexibility and optical transparency data of a celluloid plate are compatible, flexibility and combustion heat data are not. This is an important technical detail, but it does not affect the theory of knowledge, as qp-incompatibility is supposed to do. Nor is there reason for changing 'The World View of Physics' when various probabilities are found to be connected by a *law*, and when this law of 'unitary transformation' between unit magic P-squares is best expressed in terms of a geometrical law with each P having a *vector* direction in a plane structure, denoted by the complex symbols ψ. Therefore I cannot see that 'the logical(?) structure of quantum mechanics has made profound changes in our scientific thinking' or has given us 'an epistemological lesson' by 'forcing us, in the ordering of experience, to pay proper attention to the conditions of observation' (Bohr). It is obvious that the same holds for flexibility and combustion heat data. Yet their incompatibility has never made profound changes in our scientific thinking. Nor is our *thinking* changed when microphysical experiments are found to have the character of (ordinary) games of chance—except under the erroneous notion that quantal statistical distribution can never be reduced to deterministic interpretation whereas classical statistical distributions can. Actually, the difference between an ordinary game of chance and the atomic *pq*-game is technical but not essential. The *pq*-game is a physical innovation, but it does not give us an epistemological lesson.

31. *The Copenhagen language*

It was quite natural that the paradox of wave *versus* particle phenomena and the indeterministic character of the latter at first led to an emphasis on the differences between ordinary and quantum physics. But after so many years it is time to return to a less dramatic presentation of the facts, to stress the essential similarities in spite of technical differences.

The usual tendency to describe the simplest physical situations in the most devious manner may be illustrated by using the quantum language for describing the fall of stone from the original height h_0 to the ground, as follows. There is an expectation function $P(h, t)$ which is very large for $h=h_0-\frac{1}{2}gt^2$, and very small for other h-values, a so-called delta function. As long as the stone is left alone, the maximum of this function (not the object itself) travels downwards in the course of time. Suddenly the expectation function $P(h, t)$ shrinks or disappears, *either* when the stone actually arrives on the ground, *or* when the expecting observer sees it reaching the ground, *or* only half a year later when someone finds a plant marking the place of impact, the stone having carried along some plant seeds. The time moment of shrinking of the function $P(h, t)$, according to the *either-or-or* above is left open, or it may be shifted at our discretion (Heisenberg)—provided that this description makes sense at all. But will it make more sense when the stone is replaced by a particle, the delta function by one with a broader maximum, and the sprouting seed by the later development of the film?

Further developments of the Copenhagen language are even less reassuring, in particular Heisenberg's description(1) of what happens when a photon, after hitting a semi-transparent mirror, is found on the reflected side, when with the same *a priori* probability it *could* have been found on the transmitted side. 'The experiment [of ascertaining the photon] at the position of the reflected packet thus exerts a kind of action, a reduction of the wave packet, at the distant point occupied by the transmitted packet; and one sees that this action is propagated with a velocity greater than that of light.' Heisenberg then assures us that relativity theory is not violated since this 'action' cannot be used to transmit a signal. To my mind, his description is as absurd as if he would speak of two 'wave packets' of expectation running down the two sides of the blade in our ball-knife game, until the actual observation of the ball on one side produces an 'action' transmitted with superluminar velocity to make the 'wave packet' on the other side shrink suddenly. Only constant repetition may condition us into a state of mind where we shall accept such subtleties as enlightening.

It may be unfair to quote views enunciated during the first turbulent years following the establishment of quantum mechanics. But even in 1955 things have not become clearer when Heisenberg[10] repeats his story of the wave packet of expectation which suddenly shrinks by a 'transition from the possible to the actual' at the moment of an observation of a particle—or at some other moment of our own choosing (!): 'It is entirely possible to imagine this transformation, from the possible to the actual, moved to an earlier moment of time.' And why? we ask: 'For the observer does not produce the transition.' Even after trying to clarify these ideas by application to the more familiar example of the ball-knife game, I have not been able to discern an intelligible message in Heisenberg's words. Nor does he deserve a medal for clarity when he continues: 'The representation of a packet of probabilities is completely "objective", that is it does not contain features connected with the observer's knowledge. But it also is completely abstract and incomprehensible since the various mathematical expressions $\psi(q)$, $\psi(p)$, etc., do not refer to a real property; it thus, so to speak, contains no physics at all.' I have tried to understand this by applying the same words to a ball dropping along one though unpredictable side of a blade, but to no avail. At any rate, if 'objective' goes together with 'abstract', with 'incomprehensible', 'no real property', and 'no physics at all', then it is time to replace paradoxical play with philosophical vocabulary by simple language describing the comparatively simple connections between objective physical data. But I must confess that not so long ago I myself was deluded by this sophisticated quantum philosophy, many years after K. R. Popper(11) had given us a clear analysis of its allegations. For more recent criticism refer to the writings of B. Fogarasi,(12) H. Margenau,(9) and P. Feyerabend,(13) to name but a few representatives in the rising flood of opposition to the dominant doctrine.

I know, of course, that quantum theory has introduced very important innovations into the science of mechanics. The novelties do not consist, however, in the discovery that there are mutually incompatible observables as such; plenty of 'ordinary' examples can be mentioned to illustrate the same concept. Nor does the novelty consist in the realization that statistical distributions are not reducible to individual cause-effect relations; the same applies to classical games of chance. The innovations of quantum mechanics are purely physical; they do not call for a new epistemology and philosophy of knowledge, nor for a new language. They may rather be seen in the following points.

(1) There is a class of 'physical states' of a mechanical system ascertainable by tests with macroscopic instruments and reproducible by repetition of the same test, hence 'objective'. Any two of these objective physical states are connected by a definite transition probability P which is two-way symmetric with respect to the initial and final state.

(2) The transition probabilities P from state to state are interconnected by a general P-correlation law which is, and for mathematical reasons can only be, that of unitary transformation. The same law is known also as the wave-like law of probability interference via complex quantities ψ symbolizing directions in a plane. Tensors play the part of observables.

(3) Conjugate dynamical variables q and p are 'relative' with constant probability density in q-space for constant p, and vice versa. From this follows that the probability amplitude $\psi(q, p)$ must be of the complex exponential form, $\psi(q, p) = \exp(2i\pi qp/\text{const})$, that is, it must be a 'wave function' of wave length $\lambda = h/p$ if the constant is denoted as h.

These are the features which characterize quantum mechanics within the general category of games of chance. They cover all the extraordinary qualities attributed to ψ, to non-commutativity, to wave-like appearances of particles, and all the rest. Actually there is nothing extraordinary involved which would justify a new philosophy in which the 'objective' is 'abstract', 'incomprehensible' and 'contains no physics at all'. Quantum mechanics is physics; it describes regular connections between objective, that is reproducible data pertaining to 'real things', electrons, atoms, field oscillators, etc., revealed in tests with macroscopic instruments, data which are connected by statistical law. But there is no reason to disregard the standards of common sense and common language just in order to save an obsolete duality of pre-Born days combined with a subjectivistic interpretation.

All this has been said before, though mostly by philosophers of science. Yet, as a quantum theorist by profession, I must join their ranks, even after having tried for years to imbue my students with the Copenhagen spirit, but having become increasingly sceptical of my own words.

32. *Schrödinger's equation is not a process equation*

We now begin a closer analysis of a few technicalities which have, however, played a major part in long drawn-out disputes among the experts.

In the literature on the quantum theory of measurement one reads:

(*a*) A ψ-function represents one state of a system, or of an ensemble of like systems. (Correct!)

(*b*) A ψ-function represents a sequence of different states at consecutive times connected by the Schrödinger differential equation as a kind of deterministic equation of motion. (Wrong!)

Obviously (*a*) and (*b*) cannot both be true. Let us find out what the statistical interpretation has to say to this contradiction.

Statement (*a*) seems rather odd at first sight since a quantity ψ such as $\psi(S_i, A_m)$ connects *two* states, S_i and A_m, with one another. Nevertheless, the entirety of quantities $\psi(S_i, A_m)$ that is the function ψ for $m=1, 2, \ldots$ may indeed be regarded as characteristic of, or as representing, the *one* state S_i (The Hilbert vector S_i is determined by the complete set of its orthogonal components). The same state S_i can be represented just as well, however, by the function $\psi(S_i, B_j)$ for $j=1, 2, \ldots$ where $B_1 B_2 \ldots$ is another complete set of orthogonal states of the system. This viewpoint is expressed by (*a*) above where the 'state' S_i denotes a *physical state* in which the system emerges from an S-measuring device. Opinion (*b*) erroneously regards the *many* quantities $\psi(S_i, A_1)$ and $\psi(S_i, A_2)$, etc., as representing the many different states A_1 and A_2, etc.; it further maintains that the function $\psi(S_i, A_m)$ for $m=1, 2, \ldots$ represents a statistical distribution over the many states A_m, and likewise that $\psi(S_i, B_j)$ represents a distribution over the many states B_j (rather than that both functions represent one and the same state S_i as (*a*) correctly maintains). The multistate view (*b*) is inadmissible also when the states $A_1 A_2 \ldots$ are exemplified by the positional states $r_1 r_2 \ldots$ at time t_A and when the states $B_1 B_2 \ldots$ are exemplified by the positional states r_1 $r_2 \ldots$ at time t_B. According to the wrong but widely accepted view (*b*), the function $\psi(S_i; r, t_A)$ is said to represent a statistical distribution over the positions r at time t_A, and $\psi(S_i; r, t_B)$ the same at time t_B. Actually, according to (*a*), both functions of r at different times t are but different representations of the one physical state S_i. But since both functions are representations of the same physical state S_i, *Schrödinger's equation merely connects different mathematical representations of one and the same physical state.* It does not describe a physical process, neither in an individual system nor in an ensemble. Schrödinger's equation is not a process equation.

It is true that $\psi(S_i; r, t_A)$ for various values of r represents amplitudes of *expectation* that the system, which *is* in the physical state S_i, will jump to this or that physical state r if someone would carry out position measurements at time t_A. It is misleading, however, to speak of the function $\psi(S_i; r, t_A)$ as describing a *state* of expectation for various r at t_A. The term 'state', if once denoting a physical state of a system ascertained by a measuring instrument, should always be used to denote a *physical* state. As will be seen in the next section, confusion results when one uses the same word 'state' alternately for a *physical state* of a system, and for a *state of expectation* in some observer's brain. Let us keep in mind, therefore, that the functions $\psi(S_i; r, t_A)$ and $\psi(S_i; r, t_B)$ are both representations of one and the same physical state S_i.

The wrong idea that the two ψ-functions just mentioned represent two different 'states' at different times t_A and t_B may be attributed in part to the careless notation $\psi(r, t)$ rather than $\psi(S_i; r, t)$, that is omitting reference to the all-important state S_i which ψ actually represents. In Schrödinger's wave mechanics $\psi(r, t)$ was indeed thought as a physical state in space varying with time. Born first reinterpreted $\psi(r, t)$ as describing an actual distribution over the positional states r. But in the modern statistical view $\psi(r, t)$ represents *one* physical state S_i not named in the usual notation. $\psi(r, t)$ also signifies statistical information, namely betting odds that the system *would* turn to various positions r if it should be subjected at time t to an r-meter. These betting odds are sometimes misnamed as representing a 'state of expectation'; the latter concept is of course quite different from that of a 'physical state'.

33. *Theory of measurement*

According to Heisenberg and von Neumann, one may distinguish between the following three steps of a measurement:

(*a*) Before the measurement the *state* of the system is represented by a function $\psi(x, t)$. This state changes in a continuous and reversible fashion, dominated by the Schrödinger equation as a kind of *process equation.*

(*b*) During an act of measurement, the *state* undergoes a

sudden discontinuous change, known as a reduction of a wave-parcel, not controlled by the Schrödinger differential equation.

(*c*) After the measurement the *state* changes continuously again as in (*a*).

There is disagreement on whether the 'reduction of a wave-parcel' takes place with the velocity of light as a physical process, or with the velocity of thought as a mental process. There is also discussion whether the Schrödinger equation stops operating when an instrument records a datum, or only when an observer takes notice of this datum. In an effort to clarify the contraction hypothesis, Heisenberg([10]) tells us that a ψ-function, also known as a Hilbert vector,

(1) 'is completely "objective", that is no longer contains features connected with the observer's knowledge';

(2) 'completely abstract and incomprehensible, since the various expressions $\psi(x)$ and $\psi(p)$, etc., do not refer to real space or to a real property; thus, so to speak, ψ contains no physics at all';

(3) 'the discontinuous reduction of wave-packets is a transition from the possible to the actual'.

However, Heisenberg's description (*a*), (*b*), (*c*) with the 'equation of motion' on, off, and on again is so inconsistent that Schrödinger justly refuses to accept the discontinuous event (*b*) since he wishes to retain (*a*) and (*c*) in his wave mechanics. In contrast, the statistical interpretation rests on uncontrollable changes of state of a micro-physical object during a test by means of a macroscopic instrument. I therefore cannot agree with (*a*) and (*c*), even if Heisenberg asks us to accept (*a*), (*b*), (*c*) at the same time. The consistent statistical theory rather runs as follows:

(*a*′) Before the measurement is carried out, the system is in a certain initial physical state S_i after emerging in this state from a previous *S*-measurement. This state persists until a further measurement is carried out. (Indeed, what could be the meaning of a 'change of physical state' when the system is isolated, when no determination of the supposed change of state is permitted, and the system does not 'know' what kind of test it will undergo next.)

(b') During an act of measurement with an R-meter the system is compelled to jump from S_i to one of the states R_1 or R_2, etc. It may actually arrive in the state R_n; the relative frequency of this process is $P(S_i, R_n)$.

(c') After the system is separated from the R-meter, it remains in the state R_n until another test compels it to change its state once more.

Again I yield to the annoying habit of testing the highly delicate quantum logic and language by applying it to crude ordinary examples. When one is told that a balloon gradually increases its volume, then, after contact with a needle suddenly reduces its volume, and thereafter increases its volume again, then he trusts that the term 'volume' stands for the same physical concept in all three cases. If he expects the same to hold for the term 'state' in the description (a), (b), (c) of the collapsing wave-parcel, he will be disappointed. In (b) 'state' means physical state of the system as it should; in (a) and (c) it means 'state of expectation' in some observer's mind. And no pseudo-philosophical talk about the subjective character of quantum mechanics, about the transformation from the possible to the actual can conceal the fact that the usual theory of measurement uses one word in two entirely different meanings. The result has been confusion among those who 'listen to the words of the theorists rather than looking at their deeds' (Einstein). The famous 'contraction of a wave-parcel', the topic of innumerable discussions, is a fiction resting on a double meaning of the word 'state'. On the other hand, (a'), (b'), (c') is the true story as told in the spirit of the statistical interpretation. It is a simple story, with the word 'state' having but one meaning, namely, physical state.

In classical physics one is wont to fill the gaps between actually observed data by a continuity of hypothetical intermediate data, as illustrated by the path of a firefly viewed at night (Margenau). Quantum theory prefers *not* to speak of what happens to a system between two observations of its state; it ascribes changes of state only to intervention from outside in tests. For example, if a first test of a system of many particles, of a gas, displays an improbable distribution, and the next test displays a more probable distribution, then this change of physical state is ascribed to the test. It can be shown that the two-way symmetry of the transition probabilities leads to the result that a test will with great probability lead from a less to a more probable distribution (section 15).

The Heisenberg–von Neumann contraction hypothesis (*a*), (*b*), (*c*) has often been criticized severely. Margenau sees the origin of 'confusion and welter of contradictions' in a double meaning of the word 'probability', first as an objective quality describing the results of a long series of tests, and second as a degree of certainty of an observer's knowledge, so that 'one wonders whether (quantum theory) was designed to deal with the physics of particles or the psychology of perception'. My own criticism is directed against the double meaning of the word 'state'; it may corroborate Margenau's view.

It is small wonder that misunderstandings about the role of the ψ-function with its alleged sudden contraction, with its half physical, half psychological content, etc., has led to the widespread view that the quantum theory, and the complex quantity ψ in particular, is abstract, sophisticated, fictitious, unpictorial, incomprehensible, has no direct physical meaning, and yet for some unaccountable reasons of 'duality' leads to correct results. I think that the simple outline of the theory developed in the previous chapters does not support this ancient folklore any more.

Indicative of the present confusion regarding ψ is the following list of seven opinions that a ψ-function describes:

(1) The physical state of a continuous material medium in space and time, with particles as mere appearances (Schrödinger 1926).
(2) A continuous pilot wave which controls point events along its course (de Broglie).

(3) A fluid containing hypothetical quantum forces invented *ad hoc* so as to determine its own motion according to the laws of hydrodynamics (Bohm).
(4) *One* definite state, or a sequence of *many* states of an object, or of a statistical ensemble of objects (the textbooks vacillate between these opinions).
(5) A wave state in space (for example in the transmitted as well as in the reflected part of a 'wave packet' in case of partial reflection) contracting with super-luminal velocity either when a point event takes place (in one of the two parts) or only when an observer gains knowledge thereof (Heisenberg, 'waves of expectation').
(6) A mathematical symbol incapable of pictorial representation, completely abstract and containing, so to speak, no physics at all, yet completely 'objective' since not referring to any observers' knowledge (Heisenberg, the Copenhagen language).
(7) A ψ-function is a well-ordered (Chapter III) list of betting odds, based on past statistical experience, for the diverse outcomes of specific tests of a microscopic object with macroscopic instruments.

Unfortunately, the last opinion takes all the fun out of the quantum philosophy although, in my opinion, it is the only tenable one.

A great deal of confusion has its origin in the early, but wrong, analogy between matter particles and photons on the one hand, ψ-waves and electromagnetic waves on the other. There are individually identifiable real particles (producing tracks, etc.), but there are no 'photons' as real identifiable things. On the other hand, electromagnetic fields are real (kickable) things, whereas ψ-waves are tables of probabilities for events (rather than real things as Schrödinger first thought). The quantum theory can be applied to real things, to matter particles as well as to electromagnetic waves. In particular, if the motion of a particle is harmonically analysed into components of frequency ν_k, then quantum theory says that it can give out and receive energies only in amounts $h\nu_k = E' - E''$. Similarly, an electromagnetic or other field may be analysed harmonically into standing vibrations of various frequencies ν_k which change their energy only in quantized amounts $h\nu_k$ in resonance with a corresponding change of a matter particle. However, this change of field energy must not be thought to take place locally at the place of the particle, as though a 'photon' were swallowed or emitted. The change of radiation energy rather takes place in the standing wave as a whole. This at last is the view of the *quantum theory of radiation* as developed around 1930 by Fermi, Heitler, and others. There are no localized photons. And quantum theory regards the creation and annihilation of matter particles not as true creation out of nothing or annihilation into nothing, but as transitions of the particle from positive to negative energies, accord-to Dirac's hole theory; the particle thus persists as a 'thing'.

34. *Bohm's pseudo-causality*

When balls are dropped along a board spiked with nails (Galton board) they arrive at the bottom with a statistical distribution known as the Gaussian law. This law can be deduced by pure mathematical reasoning based on the assumption that each ball has an even chance of dropping to the right or left of a nail. The Gaussian distribution is thus reduced to simple fundamentals; it does not require any further explanation. However, someone proposes the following 'causal' theory. First he interprets the subsequent points of arrival of the balls as the particles of a gas, although they represent events maybe years apart. Second, he wonders why his self-invented 'gas' displays a density maximum in the centre rather than spreading out evenly, as a real gas would do. Third, he embarks on inventing a hydrostatic force distribution throughout his fancied gas adjusted in such a manner that, *if* this force existed, it would produce the same density distribution in a real gas according to the gas pressure laws which the 50:50 probability yields anyway according to Gauss. An *ad hoc* constructed hypothetical force in a non-existent 'gas' may have esthetic and even heuristic value. Regarding it as a causal explanation of Gauss's law on a deeper level of understanding would stretch the meaning of the words 'explanation' and 'understanding' beyond recognition, however.

It so happens that Bohm's theory[14] is the exact anogueal of the above 'causal explanation' of the Gaussian distribution. Bohm first reinterprets the statistical density $|\psi(x, t)|^2$ in *potential* experiments as the density of a fluid. Then wondering why his self-invented fluid is not distributed with constant density in space and time, as would an ordinary fluid in the absence of compressive forces, Bohm goes on inventing a hydrodynamic force distributed throughout his background fluid distributed in such a manner that, *if* it actually existed, it would produce the same space-time distribution according to gas dynamics which the fundamental probability metric, and the Schrödinger equation as a special application of this metric, yields anyway. Finally he would have us believe that the individual particle positions are floating along in the stream of this background fluid.

There are two fundamental objections to this theory. First, following wave mechanics, Bohm's picture suggests that $\psi(x, t)$ represents an actual statistical distribution state of particles over x at various times t. In fact, $\psi(x, t)$ stands for $\psi(S; x, t)$ indicating that the particles *are* all in the same state S, and that $|\psi(x, t)|^2$ is their probability of jumping from S to x at t if someone *should* carry out x-tests (rather than A-tests in general, x and t having no special preference in the general probability metric). At any rate $|\psi(x, t)|^2$ is a *transition* probability; it does not indicate a distribution of particles. Second, even if $|\psi(x, t)|^2$ were a statistical density distribution at various times t, according to an antiquated version of the statistical interpretation, even then would a properly adjusted 'quantum force' invented *ad hoc* hardly offer a deeper understanding of the particle behaviour. It would merely be an interesting *analogy* to the laws of hydrodynamics, but rest on no more fundamental grounds than a hydrostatic 'explanation' of the Gaussian distribution at the bottom of a Galton board.

35. *Summary*

In contrast to the positive content of previous chapters, the present chapter is a critique of current interpretations of the quantum formalism. If Born's statistical particle interpretation is accepted seriously, there is no room for 'duality' any more; wave phenomena are mere appearances of *real* particles. There are technical, but no essential, differences between 'ordinary' games of chance and quantum games. Therefore, the subtleties of the Copenhagen language with their alleged new 'epistemological lessons' are open to criticism. The subjective trend of the current quantum philosophy finds its culmination in the fairy tale of the contraction of a wave-parcel, which rests on a double meaning of the term 'state', namely, first 'physical state' as actually observed, then 'state of expectation' of an observing subject. In the consistent statistical interpretation nothing happens to the state of a system unless it is in contact with a measuring instrument. When cleared of archaic connotations, quantum theory becomes a consistent unitary particle theory in which wave-like appearances are natural consequences of elementary non-quantal postulates.

Although the developments of this book primarily envisage the quantum mechanics of matter particles, our results apply also to electromagnetic and other fields. Quantum theory decomposes fields into 'field oscillators' or other constituent parts with generalized 'co-ordinates' and 'momenta' which are subjected to the same formal rules of quantization as the co-ordinates and momenta of matter particles. And the creation and annihilation of matter particles are treated as jumps from a positive to a negative energy level or *vice versa*, again dominated by the same general probability metric.

RETROSPECT

From Dualism to Unity, from Positivism to Realism

The search for a rational interpretation of quantum phenomena with its dualistic appearances has led to the following clear-cut question: Suppose that matter consists of particles; how are we to explain their occasional wave-like behaviour? Vice versa, suppose matter consists of a continuous substratum supporting waves; how do we account for corpuscular phenomena? The question assumes of course that matter can have either the one or the other 'real' constitution. Although this assumption is most natural for the physicist, it is opposed to the doctrine that there is no absolute physical reality, that we merely construct mental pictures coloured by subjective sense impressions, and then arrange them into coherent schemes or theories. 'Describe economically, but do not search for hypothetical realities behind the phenomena' is indeed a most comfortable philosophy in view of the dilemma of two rival pictures. It is significant, however, that this detached neutrality was accepted by the theorists only *after* their efforts to establish a unitary theory of matter had failed—temporarily at least.

Neutrality toward the wave-particle paradox, elevated by Niels Bohr and Heisenberg to a 'fundamental principle' of duality inherent in all matter (and fields) is diametrically opposed to the *realism* of Einstein to whom 'the concepts of physics refer to a real world, to things which claim real existence independent of perceiving subjects'. At heart all physicists are realists in Einstein's sense to whom there could only be *either* discrete particles *or* continuous waves of matter, rather than 'whole peas and pea soup at the same time', even if the two 'pictures' might be complementary. Physical questions of the either-or type are not solved by tranquillizing pills dressed in philosophical language. Using the age-old scepticism of philosophers as to the reality of the external world to serve as a cover for our temporary ignorance and indecision, is the policy of 'if you can't explain it, call it a principle, then look down on those who still search for an explanation as unenlightened'.

Clear-cut unitary quantum theories of matter have indeed been advanced, but found wanting in one way or other. In particular I refer to Schrödinger's fascinating idea that what *appears* as a particle is *in reality* the crest of a wave group; the crest moves indeed, according to the wave equation, on the same path and with the same velocity on which a particle would move according to the laws of mechanics. This wave interpretation and explanation of corpuscular phenomena had to be abandoned, however, since wave crests rapidly level off and thus lose their corpuscular appearance. After this failure, Max Born made the opposite proposal of a unitary theory: what *appears* as a wave intensity is *in reality* a probability density of particles as the real constituents of matter. Born's unitary particle interpretation is now accepted by practically all physicists. Yet many of them in their idle hours still pay lip service to the neutrality doctrine of waves and particles on an even level—among them, strangely enough, Born himself, and not without a certain (negative) ideological reason. For whereas Schrödinger's wave crest interpretation plus explanation of particle appearances broke down with the wave crests themselves, the particle interpreters have never felt any need for giving us an explanation as to why particles ought to obey wave-like laws in their statistical behaviour; they follow Niels Bohr in believing that wave appearances of particles are manifestations of a fundamental irreducible trait of nature displaying particle as well as wave features in a complementary fashion, without permitting a decision in favour of one or the the other 'picture'. In my opinion, the temporary impasse between two theories of the real constitution of matter, either real particles with wave appearances, or vice versa, has little to do with the deep philosophical problem of reality which has intrigued thinkers from the Greeks via Descartes to modern times. Appealing from the wave-particle controversy to the philosophy of knowledge and engaging the whole armoury of positivism to convince us that we here are faced only with a pseudo-problem, is turning the vice of ignorance into a virtue. It is an appeal to a metaphysical doctrine which provides a name to those paradoxical quantum features which call for an explanation, that is, for a better understanding on the grounds of simple *non-quantal* principles of a familiar character. It is true that all 'principles' are, in the last resort, metaphysical; but the non-quantal ones have at least the advantage of not possessing the *ad hoc* character of their quantum equivalents.

This study poses, and I hope answers, the following questions: Are there elementary reasons for

(*a*) The lack of sufficient causation, that is for the statistical character of atomic events?

(*b*) The domination of statistical laws of particle mechanics by a wave-like interference of probabilities via a complex quantity ψ?

(*c*) The periodic connection between co-ordinates and momenta, known from the rules $E = h\nu$ and $p = h/\lambda$ and their modern counterparts?

The following answers have been given in the preceding chapters:

(*a*) *Determinism or Chance Law.* Statistically ruled events, in ordinary games of chance, in automobile accidents, as well as in quantum tests violate the law of sufficient causation. This law, which is disproved by common experience, must not be confused with the law of sufficient reason. Saying: 'I cannot *think* of anything happening without a specific cause', does not mean much. On the contrary, there is *sufficient reason for insufficient causation.* The empirical law of *cause-effect continuity* leads, on grounds of simple reasoning to the conclusion that there must *also* be discontinuous acausal events controlled by statistical law. Experience alone is to decide which of the two conflicting empirical laws, determinism or continuity, is correct; and experience decides in favour of cause-effect continuity, *hence* against the possibility of restoring strict determinism for individual events in statistical ensembles. Determinism is preserved, however, in the diluted form of statistical law. The hope of finding 'hidden causes' can thus be repudiated by positive arguments rather than by relying on negative evidence alone.

(*b*) *Wave-Particle Dualism.* The quantum experts tell us that matter is dominated by a dualistic contrast between particle and wave qualities as a deep-seated feature of the microphysical world, attenuated by the fact that there is no direct contradiction between the two since under the uncertainty principle the two theories must remain incomplete. They are *complementary* to one another, and various phenomena (such as the Compton effect, the diffraction of electronic rays through crystals, etc.) may be interpreted either by wave interference or by particle mechanics, both within the limits of uncertainty. The question remains, however, whether wave-particle duality and complementarity are to be regarded as novel and independent features of the microphysical world, and thus are of fundamental physical and philosophical importance, or whether they are but superficial appearances deducible from a unified basis. Actually, Max Born's statistical interpretation has destroyed the root of the dualistic view: particles alone are the *substance* of matter, and Schrödinger's wave intensity is only one of many *qualities* belonging to particles, namely the statistical density distribution of particles in a series of equal experimental arrangements. Contrasting a thing to one of its many qualities is illogical; there is no duality any more.

Yet even Born's clarification has left us with the strange fact that particles, for reasons unknown, obey wave laws of interference under a wave equation, as though a mysterious power compelled them to maintain at least the *appearance* of duality. This inexplicable obedience to wave laws has gradually been elevated to an immanent quality. The term 'immanent' in this connection signifies a quality originally viewed as quite surprising, yet gradually become so familiar that it is now regarded as inseparably *belonging* to an object. Typical examples of immanent qualities are, or were: the horror vacui, the aim of a loadstone to attract iron, and last not least, the universal attraction of celestial and terrestrial bodies. These qualities were first regarded as implanted into matter by an unfathomable whim of Nature, later regarded as 'inherent'. The wave-like particle qualities are of the same category: first they were considered as oddities and, after many futile efforts, as inexplicable. Later they became familiar, and are now incorporated into the theory as integral parts, not to be questioned any more, as inviolable articles of faith in the cult of duality and complementarity. But similar to the belief in gravity as 'belonging' to matter, the acceptance of duality as 'ultimate' has blocked for many years all further progress toward a deeper understanding of microphysical law from a unified point of view. Actually, the wave-like interference of probabilities is not an independent quality imposed on particles by some inscrutable power. It rather is a direct consequence of the elementary proposition that there is a general correlation law between the probabilities of transition from state to state. For purely mathematical reasons this law is that of 'unitary transformation', identical with that of ψ-interference. This law, then, is of an almost *aprioristic* character, rather than an *ad hoc* assumption which we have to make because it works so well. By the way, there is a close analogy of the P- and ψ-structure with geometrical structures in a *plane*; each complex quantity ψ represents a vector in a plane, giving direction to the corresponding transition probability.

(*c*) *Quantum Periodicity*. Long experience coupled with mathematical ingenuity has led to the result that the probability amplitude $\psi(q, p)$ has the form of a complex imaginary function, $\exp(2i\pi qp/\text{const})$, periodic in p for given q, and *vice versa*, the constant having the dimension of an action; it is known as Planck's constant. The periodicity of the function $\psi(q, p)$ is equivalent to Born's commutation rule and Schrödinger's p-operator rule, usually regarded as another immanent quality of the microphysical world, as a novel and unexplainable 'quantum feature'. It was seen, however, that the quantum periodicity is by no means inexplicable, nor a manifestation of Nature's inscrutable desire to satisfy Bohr's 'principle' of duality and complementarity. It rather is a consequence of the general probability metric, that is ψ-interference in combination with the familiar non-quantal feature of linear co-ordinates q and momenta p to have physical significance only up to additional constants, and yielding a *constant* statistical density in qp-space, as known from classical statistical mechanics. The wave-like laws of quantum dynamics are thus explained on the basis of *unitary particle theory*.

The considerations (*a*) may strengthen the ideological substructure of quantum mechanics as a *statistical* theory against the defenders of determinism; (*b*) and (*c*) may dispose of the belief in an inscrutable 'principle' of duality. The Bohr-Heisenberg idea of two pictures of equal rank, neither of them presenting physical reality, can be overcome by the realistic world view of Einstein that there are real things (particles) with objective, that is, reproducible qualities.

It is instructive here to look at the history of dualism since 1923 in retrospect.

(1) Dualism, initiated by Einstein's working hypothesis of photons in 1905, began to be taken seriously when Duane in 1923 established a corpuscular mechanical theory of X-ray diffraction, and when de Broglie in 1924 interpreted discrete particle orbits in terms of waves. Dualism of matter seemed to be supported experimentally by interference patterns of electronic rays through crystals, as opposed to discrete tracks in cloud chambers and films. The simple *wavicle dualism* of 'matter particles' versus 'matter waves' was shattered, however, when better observation methods revealed that those interference patterns are produced by individual electronic impacts in a statistical manner, in agreement with Born's unitary particle interpretation.

(2) Dualism was revived, however, when Klein, Jordan, and

Wigner in 1928 showed that the usual statistical quantum mechanics for particles can be transformed into an *equivalent* formalism suggesting a continuous medium in 3-space with properly adjusted quantum features so as to yield corpuscular results. We thus were confronted with a *neo-dualism* of 'particles displaying wave-like statistical laws' versus 'a wave-bearing continuum with corpuscular-like qualities'. I strongly object to taking this observational equivalence of K-J-W's theory as a physically significant argument for dualism. First of all, the usual quantum mechanics of particles is *directly* supported by discrete tracks and discrete impacts on interference patterns. Second, the basic quantum rules for particles are *simple.* Third, they can be *explained* by reduction to simple and plausible general postulates of symmetry, etc. for particle physics. In contrast, the quantum rules needed for K-J-W's fluid are mathematically involved (many quote their papers, few have read them). Second, they are physically devious (when I see tracks and discrete impacts I will ascribe them to discrete particles, rather than taking them as manifestations of a hypothetical medium supporting a non-linear wave equation for non-commutative wave functions). And third, those anomalous qualities of K-J-W's fluid cannot be justified by reduction to simple and general postulates for a fluid; they are to be assumed in order to yield the desired equivalence safeguarding a modicum of duality. (Similarly, in order to save the idea of a 'medium' one could ascribe Michelson's result to a hidden *ether* endowed with such qualities as to be *equivalent* with Einstein's simple invariance postulates.) Furthermore, other equivalent quantum schemes have been devised by de Broglie, Bohm, Bopp, and others, so that one might as well speak of multiplicity rather than duality, if these theories did not suffer from the same defect as the continuum theory of K-J-W, namely of being *ad hoc* devised with the aim of satisfying this or that ideology, rather than giving a simple and immediate account of experience.

(3) When seeing that the construction of observationally

equivalent theories is of little physical significance, quantum theorists now insist that there still is duality in the opposition of 'particle position' versus 'particle momentum' separated by uncertainty. This, however, at best is *neo-neo-dualism.* In fact it is the unitary quantum mechanics of particles with wave-like statistical laws. Still waving the flag of duality in this context is an empty ritual dictated more by sentimental motives than by a desire for clarity. It is reminiscent of the fancy use of the word 'democracy'.

In its critical Part V this study is directed against the ideological and linguistic excesses of certain dualists. There is no doubt of course that the wave-particle contrast is of great heuristic value. My only objection is that the quantum translation rules do not in themselves represent the ultimate bottom of insight. There are deeper reasons for their validity. And when Bohr speaks of 'the dilemma regarding the corpuscular and wave properties of electrons', I hold that *there is no dilemma* when one learns that the wave-like phenomena of particles are not anomalies, but are exactly what one ought to expect of particles under simple postulates of symmetry, invariance, etc. Philosophically speaking, I think that even in quantum theory one can return from the dialectical positivism of opposing pictures to plain ontological materialism which from Galileo to Einstein has been the ideological background of natural science.

APPENDIX I

SURVEY OF ELEMENTARY POSTULATES

(*a*) The postulate of *continuity* maintains that 'in a case of deterministic cause-effect relations, an infinitely small change of cause will never produce a finite change of effect'. It leads by necessity to the admission of discontinuous acausal events controlled by statistical law.

(*b*) When the continuity postulate is applied to the separability of various 'states' by means of semi-permeable instruments, and is supplemented by the postulate of *reproducibility* or *confirmability*, then it leads to a 'splitting effect' accompanied by transitions from one state to other states in tests, controlled by statistical laws.

(*c*) The postulate of *symmetry* for the transition probability between any two states corresponds to the classical reversibility of mechanical processes; it entails that the tables for the transition probabilities from one set of orthogonal states to another set must be unit magic square matrices.

(*d*) The postulate: 'there ought to be a *general metric law* connecting the various probability tables' is satisfied only (?) by the metric of 'unitary transformation'. The latter, however, is identical with the superposition law for probability amplitudes.

(*e*) The postulate that observables depend on q- and p-values only up to additional arbitrary constants, together with the postulate of a *constant statistical density* in q-space and in p-space, taken over from classical statistical mechanics, is the postulate of *invariance* with respect to Galileo and/or Lorentz transformation. It entails that the amplitude $\psi(q, p)$ must necessarily have the periodic form of a complex-imaginary exponential function, $\psi(q, p) = \exp(2i\pi qp/\text{const})$. With the constant called h this is the basic formula of quantum dynamics, equivalent to the Born commutation rule and the Schrödinger p-operator rule.

APPENDIX 2

TWO PROBLEMS OF UNIQUENESS

The two uniqueness problems of Chapter III formulated in purely mathematical terms read as follows:

PROBLEM I

An infinite set of letters A, B, C, ... is used to label positive quantities

$$Q_{AA}, Q_{AB}, Q_{AC}, \ldots Q_{BA}, Q_{BB}, Q_{BC}, \ldots$$

whereby

$$Q_{AB} = Q_{BA} \text{ and } Q_{AA} = 0 \tag{1}$$

Find a mutual correlation law between the Q's, symmetric in the letters. Notice that the desired law may either be ***trigonal***, so as to connect three Q's, for example, Q_{AB}, Q_{BC}, Q_{AC} labelled with three letters, so that one of the three Q's is uniquely determined by the two others, for any choice of three letters; or *tetragonal*, so as to connect six Q's, for example, Q_{AB}, Q_{AC}, Q_{AD}, Q_{BC}, Q_{BD}, Q_{CD} labelled with four letters, so that one of the six Q's is uniquely determined by the five others, for any choice of four letters, or *pentagonal*, . . . *polygonal*.

A solution of the problem is obtained from the geometrical model of the letters A, B, C, . . . representing points, with L_{AB} . . their distances, $Q_{AB}=f(L_{AB})$ as distorted distances, and $\phi_{AB}=\phi_{BA}$ associated vectors $|\phi|=L$, the desired law being

$$\phi_{AC} = \phi_{AB} + \phi_{BC},\ L = |\phi|, \text{ and } Q = f(L) \tag{2}$$

If the points are in 1 (2, 3, . . .) dimensional space, then a set of 4 (5, 6, . . .) points are connected by 6 (10, 15, . . .) distances L, whereby one L is uniquely determined by the 5 (9, 14, . . .) other L's of the same group, thus yielding a tetragonal (pentagonal, . . .) Q-law.

Question: Do the Q-relations based on this model (2) represent the *only* possible general correlation law between positive quantities Q?

PROBLEM II

An infinite set of M letter groups, $A_1, A_2 \ldots A_M$ and $B_1, B_2 \ldots B_M$, and so forth are used as labels for the elements of $M \times M$ matrices: (P_{AA}), (P_{AB}), (P_{AC}), . . . (P_{BA}), (P_{BB}), (P_{BC}), . . . where

$$(P_{AB}) = \begin{bmatrix} P(A_1B_1) & P(A_1B_2) & \ldots & P(A_1B_M) \\ P(A_2B_1) & P(A_2B_2) & \ldots & P(A_2B_M) \\ \ldots & \ldots & & \ldots \end{bmatrix}$$

The elements are all positive, between 0 and 1, and satisfy:

$$P(A_kB_j) = P(B_jA_k) \text{ as well as } P(A_kA_{k'}) = \delta_{kk'} \tag{3}$$

so that (P_{AA}) is the unit matrix. Furthermore, there is to be

$$\Sigma_j\, P(A_kB_j) = 1 \text{ and } \Sigma_k\, P(A_kB_j) = 1 \tag{4}$$

that is, the matrices are 'unit magic squares'.

Find a mutual correlation law between the P-matrices, symmetric in the letters, which leaves the unit magic square quality invariant.

A solution of this problem is known as *unitary transformation*: Associate with every $P(A_kB_j) = P(B_jA_k)$ two mutually complex quantities

$$\psi(A_kB_j) = \psi^*(B_jA_k), \text{ with } P = |\psi|^2 \tag{5}$$

representing two vectors in a plane. Then give to the ψ's such phase angles (directions in the plane) that the ψ-matrices will satisfy the matrix multiplication law

$$(\psi_{AC}) = (\psi_{AB})\,(\psi_{BC})$$

Notice that this law, similar to the pentagonal law of geometry in a plane, involves a pentagonal relation law between ten P-matrices connecting five letter groups (for any multiplicity

M) so that one *P*-matrix is uniquely determined by the nine others.

Question: Is unitary transformation the *only* possible correlation law connecting *P*-matrices, leaving their unit magic square quality invariant?

REFERENCES

INTRODUCTION

1. Sir James Jeans, *Presidential Address*, *Nature*, September 1934.
2. C. F. von Weizsäcker, *The World View of Physics*, University of Chicago Press, 1957.
3. N. Bohr, *Discussion with Einstein*, Library of Living Philosophers, Vol. 7, 1949.
4. L. Rosenfeld, 'The Strife about Complementarity', *Science Progress*, **163**, 393, 1953.
5. M. Bunge, *Brit. J. Phil. Sci.*, **6**, 1 and 141, 1955.

CHAPTER I

1. K. R. Popper, 'Indeterminism in Quantum Physics and in Classical Physics', *Brit. J. Phil. Sci.*, **1**, Nos. 2 and 3.
2. E. Nagel in *Determinism and Freedom*, edited by S. Hook, New York University Press, 1958, p. 183.
3. M. Black in *Determinism and Freedom* (reference 2), p. 15.
4. C. von Weizsäcker, *The World View of Physics*, University of Chicago Press, 1949.
5. H. Margenau, *The Nature of Physical Reality*, McGraw-Hill, New York, 1950.
6. N. Bohr, *Atomic Physics and Human Knowledge*, John Wiley, New York, 1958.
7. A. Landé, 'Determinism versus Continuity in Modern Science', *Mind*, **67**, 174, 1958.
8. A. Landé, *Naturwissenschaften*, **17**, 634, 1929.
9. A. Landé, *Foundations of Quantum Theory*, Yale University Press, 1955.
10. J. von Neumann, *Mathematische Grundlagen der Quantenmechanik*, Springer, Berlin, 1932.
11. L. de Broglie, *La Physique quantique, restera-t-elle indéterministe ?*, Gauthier-Villars, Paris, 1953.
12. P. Feyerabend, 'Zur Quantentheorie der Messung', *Z. Phys.*, **148**, 551, 1957. G. Schultz, *Ann. Phys.*, **3**, 94, 1959.
13. K. R. Popper, *The Logic of Scientific Discovery*, Basic Books, New York, 1959. Also reference 1 above.
14. N. Bohr in *Albert Einstein, Philosopher-Scientist*, Library of Living Philosophers, Evanston, Illinois, 1949.
15. P. Bridgman, *Sci. Mon., N.Y.*, **79**, 32, 1954.
16. H. Reichenbach in *Philosophic Foundations of Quantum Mechanics*, University of California Press, Berkeley, 1944.

CHAPTER II

1. A. Landé, *Foundations of Quantum Theory*, Yale University Press, 1955.
2. P. A. M. Dirac, *The Principles of Quantum Mechanics*, Clarendon Press, Oxford, 1947. (Third edition.)
3. E. Schrödinger, *Brit. J. Phil. Sci.*, **3**, 109, 1952.
4. N. Bohr, *Dialectia*, **2**, 312, 1948.
5. A. Landé, *Philosophy of Science*, **20**, 101, 1953. *Amer. Scient.*, **41**, 439, 1953. Also reference 1.

6. J. von Neumann, *Mathematische Grundlagen der Quantenmechanik*, Springer, Berlin, 1930.
7. A. Grünbaum, *Time and Entropy*, *Amer. Scient.*, **43**, 550, 1955.

CHAPTER III

1. N. Bohr, *Atomic Physics and Human Knowledge*, John Wiley, New York, 1958.
2. P. Bridgman, *Sci. Mon., N.Y.*, **79**, 32, 1954.
3. W. Heisenberg in *Niels Bohr and the Development of Physics*, Pergamon Press, London, 1955.
4. Cornelius Lanczos, *Amer. Scient.*, **47**, 41, 1959.
5. J. Lennard-Jones, *Sci. Mon., N.Y.*, **80**, 175, 1955.
6. N. Bohr, *Dialectica*, **2**, 312, 1948.
7. P. Jordan, 'Quantenlogik und das Kommutative Gesetz', in *The Axiomatic Method*, North-Holland Publ. Company, 1959.

CHAPTER IV

1. C. von Weizsäcker in *The World View of Physics*, University of Chicago Press, 1957, p. 75.
2. N. Bohr in *Atomic Physics and Human Knowledge*, Wiley, New York, 1958, p. 38.
3. W. Duane, *Proc. Nat. Acad. Sci., Wash.*, **9**, 158, 1923.
4. P. P. Ewald, *Kristalle und Röntgenstrahlen, Springer, Berlin*, 1923.
5. P. Ehrenfest and P. Epstein, *Proc. Nat. Acad. Sci., Wash.*, **10**, 133, 1924; **13**, 400, 1927.
6. E. Schrödinger, *Ann. Phys.*, **82**, 257, 1927.
7. E. Schrödinger, *Phys. Z.*, **23**, 301, 1923.

CHAPTER V

1. W. Heisenberg, *The Physical Principles of Quantum Theory*, Chicago University Press, 1930.
2. Sir James Jeans, *Nature*, London, 8 September 1934.
3. K. von Weizsäcker, *The World View of Physics*, Chicago University Press, 1952.
4. R. Havemann, *Phys. Bl.*, **13**, 298, 1957.
5. M. Born, *Philosoph. Quarterly*, **3**, 139, 1953.
6. N. Bohr, *Atomic Physics and Human Knowledge*, Wiley, New York, 1958.
7. Jean-Paul Sartre, *L'Etre et le Néant*, Librairie Gallimard, Paris, 1943.
8. K. R. Popper, 'Philosophy of Science, a Personal Report', *Brit. Philosophy in Mid-Century*.
9. H. Margenau, *Phil. Sci.*, **25**, 23, 1958.
10. W. Heisenberg in *Niels Bohr and the Development of Physics*, Pergamon Press, London, 1955.
11. K. R. Popper, *The Logic of Scientific Discovery*, Basic Books Inc., New York, 1959.
12. B. Fogarasi, *Dtsch. Z. für Philos.* **3**, 190, 1957.
13. P. Feyerabend, *Proc. Aristotel. Soc.*, **32**, 75, 1958. *Z. Phys.*, **148**, 551, 1957.
14. D. Bohm, *Phys. Rev.*, **85**, 166, 1952. Refer also to H. Freistadt, *Nuovo Cimento*, Suppl. Vol. 5, Series 10, 1957.

INDEX

Printed in the United States
By Bookmasters